P9-CJW-047

# A
# TROUBLESOME
# INHERITANCE

# A

# TROUBLESOME

# INHERITANCE

*Genes, Race and Human History*

————◆————

NICHOLAS WADE

THE PENGUIN PRESS

NEW YORK

2014

THE PENGUIN PRESS
Published by the Penguin Group
Penguin Group (USA) LLC
375 Hudson Street
New York, New York 10014

USA · Canada · UK · Ireland · Australia
New Zealand · India · South Africa · China
penguin.com
A Penguin Random House Company

First published by The Penguin Press,
a member of Penguin Group (USA) LLC, 2014

Copyright © 2014 by Nicholas Wade
Penguin supports copyright. Copyright fuels creativity, encourages diverse voices,
promotes free speech, and creates a vibrant culture. Thank you for buying an
authorized edition of this book and for complying with copyright laws
by not reproducing, scanning, or distributing any part of it in any form
without permission. You are supporting writers and allowing
Penguin to continue to publish books for every reader.

Illustration credits
Page 52: Reprinted by permission from Macmillan Publishers Ltd: *Nature Genetics,*
Albano Beja-Pereira, Gordon Luikart, Phillip R. England, Daniel G. Bradley, Oliver C.
Jann, Giorgio Bertorelle, Andrew T. Chamberlain, Telmo P. Nunes, Stoitcho
Metodiev, Nuno Ferrand, Georg Erhardt. "Gene-culture coevolution between cattle
milk protein genes and human lactase genes," 35.4(2003) 311–313, copyright 2003.
91: Source: Benjamin Voight et al., "A Map of Recent Positive Selection in the Hu-
man Genome," *PLoS Biology* 4, 446–458 (2006). 132, 133, 135, 136, 137: Clark,
Gregory, *A Farewell to Alms.* © 2007 by Princeton University Press. Reprinted by
permission of Princeton University Press.

LIBRARY OF CONGRESS CATALOGING-IN-PUBLICATION DATA
Wade, Nicholas
A troublesome inheritance : genes, race and the rise of the West / Nicholas Wade.
pages cm
Includes bibliographical references and index.
ISBN 978-1-59420-446-3
ISBN 978-1-59420-623-8 (export edition)
1. Human evolution.    2. Sociobiology.    3. Race.
4. Civilization, Western.    I. Title.
GN365.9.W33 2014
599.93'8—dc23
2013040002

Printed in the United States of America
1  3  5  7  9  10  8  6  4  2

DESIGNED BY AMANDA DEWEY

# CONTENTS

# A
# TROUBLESOME
# INHERITANCE

# 1

---

# EVOLUTION, RACE
# AND HISTORY

Since the decoding of the human genome in 2003, a sharp new light has been shed on human evolution, raising many interesting but awkward questions.

It is now beyond doubt that human evolution is a continuous process that has proceeded vigorously within the past 30,000 years and almost certainly—though very recent evolution is hard to measure—throughout the historical period and up until the present day. It would be of the greatest interest to know how people have evolved in recent times and to reconstruct the fingerprints of natural selection as it molded and reworked the genetic clay. Any degree of evolution in social behavior found to have taken place during historical times could help explain significant features of today's world.

But the exploration and discussion of these issues is complicated by the fact of race. Ever since the first modern humans dispersed from the ancestral homeland in northeast Africa some 50,000 years ago, the populations on each continent have evolved largely independently of one another as each adapted to its own regional environment. Under these various local pressures, there developed the major races of humankind, those of Africans, East Asians and Europeans, as well as many smaller groups.

Because of these divisions in the human population, anyone interested in recent human evolution is almost inevitably studying human races, whether they wish to or not. Scientific inquiry thus runs into potential conflict with the public policy interest of not generating possibly invidious comparisons that might foment racism. Several of the intellectual barriers erected many years ago to combat racism now stand in the way of studying the recent evolutionary past. These include the assumption that there has been no recent human evolution and the assertion that races do not exist.

## The New View of Human Evolution

New analyses of the human genome establish that human evolution has been recent, copious and regional. Biologists scanning the genome for evidence of natural selection have detected signals of many genes that have been favored by natural selection in the recent evolutionary past. No less than 14% of the human genome, according to one estimate, has changed under this recent evolutionary pressure.[1] Most of these signals of natural selection date from 30,000 to 5,000 years ago, just an eyeblink in evolution's 3 billion year timescale.

Natural selection has continued to mold the human genome, doubtless up until the present day, although the signals of evolution within the past few hundred or thousand years are harder to pick up unless the force of selection has been extremely strong. One of the most recent known dates at which a human gene has been changed by evolution is from 3,000 years ago, when Tibetans evolved a genetic variant that lets them live at high altitude.[2]

Several instances have now come to light of natural selection shaping human traits within just the past few hundred years. Under the pressure of selection, for example, the age of first reproduction among women born between 1799 and 1940 on L'Isle-aux-Coudres, an island in the Saint Lawrence River near Quebec, fell from 26 to 22 years, according to researchers who were able to study an unusually complete record of marriages, births and deaths in the island's parish records.[3]

The researchers argue that other possible effects, like better nutrition, can be ruled out as explanations, and note that the tendency to give birth at a younger age appeared to be heritable, confirming that a genetic change had taken place. "Our study supports the idea that humans are still evolving," they write. "It also demonstrates that microevolution is detectable over just a few generations in a long-lived species."

Another source of evidence for very recent human evolution is that of the multigenerational surveys conducted for medical reasons, like the Framingham Heart Study. Borrowing statistical methods developed by evolutionary biologists for measuring natural selection, physicians have recently been able to tease out certain bodily changes that are under evolutionary pressure in these large patient populations. The traits include age at first reproduction, which is decreasing in modern societies, and age at menopause, which is increasing. The traits are of no particular importance in themselves and have been measured just because the relevant data were collected by the physicians who designed the studies. But the statistics suggest that the traits are inherited, and if so, they are evidence of evolution at work in present-day populations. "The evidence strongly suggests that we are evolving and that our nature is dynamic, not static," a Yale biologist, Stephen Stearns, concludes in summarizing 14 recent studies that measured evolutionary change in living populations.[4]

Human evolution has not only been recent and extensive; it has also been regional. The period of 30,000 to 5,000 years ago, from which signals of recent natural selection can be detected, occurred after the splitting of the three major races, and so represents selection that has occurred largely independently within each race. The three principal races are Africans (those who live south of the Sahara), East Asians (Chinese, Japanese and Koreans) and Caucasians (Europeans and the peoples of the Near East and the Indian subcontinent). In each of these races, a different set of genes has been changed by natural selection, as is described further in chapter 5. This is just what would be expected for populations that had to adapt to different challenges on each continent. The genes specially affected by natural selection control not only expected traits like skin color and nutritional metabolism but also some aspects of brain function, although in ways that are not yet understood.

Analysis of genomes from around the world establishes that there is indeed a biological reality to race, despite the official statements to the contrary of leading social science organizations. A longer discussion of this issue is offered in chapter 5, but an illustration of the point is the fact that with mixed-race populations, such as African Americans, geneticists can now track along an individual's genome and assign each segment to an African or European ancestor, an exercise that would be impossible if race did not have some basis in biological reality.

The fact that human evolution has been recent, copious and regional is not widely recognized, even though it has now been reported by many articles in the literature of genetics. The reason is in part that the knowledge is so new and in part because it raises awkward challenges to deeply held conventional wisdom.

# The Social Science Creed and Evolution

It has long been convenient for social scientists to assume that human evolution ground to a halt in the distant past, perhaps when people first learned to put a roof over their heads and to protect themselves from the hostile forces of nature. Evolutionary psychologists teach that the human mind is adapted to the conditions that prevailed at the end of the last age, some 10,000 years ago. Historians, economists, anthropologists and sociologists assume there has been no change in innate human behavior during the historical period.

This belief in the recent suspension of evolution, at least for people, is shared by the major associations of social scientists, which assert that race does not even exist, at least in the biological sense. "Race is a recent human invention," proclaims the American Anthropological Association. "Race is about culture, not biology."[5] A recent book published by the association states that "Race is not real in the way we think of it: as deep, primordial, and biological. Rather it is a foundational idea with devastating consequences because we, through our history and culture, made it so."[6]

The commonsense conclusion—that race is both a biological reality and a politically fraught idea with sometimes pernicious consequences—has also eluded the American Sociological Association. The group states that "race is a social construct" and warns "of the danger of contributing to the popular conception of race as biological."[7]

The social scientists' official view of race is designed to support the political view that genetics cannot possibly be the reason why human societies differ—the answer must lie exclusively in differing human cultures and the environment that produced them. The social anthropologist Franz Boas established the doctrine that human behavior is shaped only by culture and that no culture is superior to any other. From this point of view it follows that all humans are essentially interchangeable apart from their cultures, and that more complex societies owe their greater strength or prosperity solely to fortunate accidents such as that of geography.

The recent discoveries that human evolution has been recent, copious and regional severely undercut the social scientists' official view of the world because they establish that genetics may have played a possibly substantial role alongside culture in shaping the differences between human populations. Why then do many researchers still cling to the notion that culture alone is the only possible explanation for the differences between human societies?

One reason is, of course, the understandable fear that exploration of racial differences will give support to racism, a question addressed below. Another is the inherent inertia of the academic world. University researchers do not act independently but rather as communities of scholars who constantly check and approve one another's work. This is especially so in science, where grant applications must be approved by a panel of peers, and publications submitted to the scrutiny of editors and reviewers. The high advantage of this process is that the statements scholars make in public are usually a lot more than their own opinion—they are the certified knowledge of a community of experts.

But a drawback of the system is its occasional drift toward extreme conservatism. Researchers get attached to the view of their field they grew up with and, as they grow older, they may gain the influence to thwart change. For 50 years after it was first proposed, leading geophysicists strenuously resisted the idea that the continents have drifted across the face of the globe. "Knowledge advances, funeral by funeral," the economist Paul Samuelson once observed.

Another kind of flaw occurs when universities allow a whole field of scholars to drift politically to the left or to the right. Either direction is equally injurious to the truth, but at present most university departments lean strongly to the left. Any researcher who even discusses issues politically offensive to the left runs the risk of antagonizing the professional colleagues who must approve his requests for government funds and review his articles for publication. Self-censorship is the frequent response, especially in anything to do with the recent differential evolution of the human population. It takes only a few vigilantes to cow the whole campus. The result is that researchers at present routinely ignore the biology of race, or tiptoe around the subject, lest they be accused of racism by their academic rivals and see their careers destroyed.

Resistance to the idea that human evolution is recent, copious and regional is unlikely to vanish unless scholars can be persuaded that exploration of the recent evolutionary past will not lead to a resurgence of racism. In fact, such a resurgence seems most unlikely, for the following reasons.

## Genomics and Racial Differences

In the first place, opposition to racism is now well entrenched, at least in the Western world. It is hard to conceive of any circumstance that would reverse or weaken this judgment, particularly any scientific evidence. Racism and discrimination are wrong as a matter of principle, not of science. Science is about what is, not what ought to be. Its shifting sands do not support values, so it is foolish to place them there.

Academics, who are obsessed with intelligence, fear the discovery of a gene that will prove one major race is more intelligent than another. But that is unlikely to happen anytime soon. Although intelligence has a genetic basis, no genetic variants that enhance intelligence have yet been found. The reason, almost certainly, is that there are a great many such genes, each of which has too small an effect to be detectable with present methods.[8] If researchers should one day find a gene that enhances intelligence in East Asians, say, they can hardly argue on that basis that East Asians are more intelligent than other races, because hundreds of similar genes remain to be discovered in Europeans and Africans.

Even if all the intelligence-enhancing variants in each race had been identified, no one would try to compute intelligence on the basis of genetic information: it would be far easier just to apply an intelligence test. But IQ tests already exist, for what they may be worth.

Even if it were proved that one race were genetically more intelligent than another, what consequence would follow? In fact, not much of one. East Asians score around 105 on intelligence tests, an average above that of Europeans, whose score is 100. A higher IQ score doesn't make East Asians morally superior to other races. East Asian societies have many virtues but are not necessarily more successful than European societies in meeting their members' needs.

The notion that any race has the right to dominate others or is superior in any absolute sense can be firmly rejected as a matter of principle and, being rooted in principle, is unassailable by science. Nonetheless, races being different, it is inevitable that science will establish relative advantages in some traits. Because of genetic variants, Tibetans and Andean highlanders are better than others at living at high altitudes. At every Olympic games since 1980, every finalist in the men's 100-meter race has had West African ancestry.[9] It would be no surprise if some genetic factor were found to contribute to such athleticism.

Study of the genetics of race will inevitably reveal differences, some of which will show, for those who may be interested, that one race has some slight edge over another in a specified trait. But this kind of inquiry will also establish a wider and more important truth, that all differences between races are variations on a common theme.

To discover that genetics plays some role in the differences between the major human societies does not mean that that role is dominant. Genes do not determine human behavior; they merely predispose people to act in certain ways. Genes explain a lot, probably far more than is at present understood or acknowledged. But their influence in most situations is or can be overwhelmed by learned behavior, or culture. To say that genes explain everything about human social behavior would be as absurd as to assume that they explain nothing.

Social scientists often write as if they believe that culture explains everything and race nothing, and that all cultures are of equal value. The emerging truth is more complicated. Human nature is very similar throughout the world. But though people are much the same, their societies differ greatly in their structure, their institutions and their achievements. Contrary to the central belief of multiculturalists, Western culture has achieved far more than other cultures in many significant spheres and has done so because Europeans, probably for reasons of both evolution and history, have been able to create open and innovative societies, starkly different from the default human arrangements of tribalism or autocracy. People being so similar, no one has the right or reason to assert superiority over a person of a different race. But some societies have achieved much more than others, perhaps through minor differences in social behavior. A question to be explored below is whether such differences have been shaped by evolution.

## Social Behavior and History

The purpose of the pages that follow is to demystify the genetic basis of race and to ask what recent human evolution reveals about history and the nature of human societies. If fear of racism can be overcome sufficiently for researchers to accept that human evolution has been recent, copious and regional, a number of critical issues in history and economics may be laid open for exploration. Race may be a troublesome inheritance, but better to explore and understand its bearing on human nature and history than to pretend for reasons of political convenience that it has no evolutionary basis.

It's social behavior that is of relevance for understanding pivotal—and otherwise imperfectly explained—events in history and economics. Although the emotional and intellectual differences between the world's peoples as individuals are slight enough, even a small shift in social behavior can generate a very different kind of society. Tribal societies, for instance, are organized on the basis of kinship and differ from modern states chiefly in that people's radius of trust does not extend too far beyond the family and tribe. But in this small variation is rooted the vast difference in political and economic structures between tribal and modern societies. Variations in another genetically based behavior, the readiness to punish those who violate social rules, may explain why some societies are more conformist than others.

Social structure is the point at which human evolution intersects with history. Vast changes have occurred in human social structure in all three major races within the past 15,000 years. That is the period in which people first started to switch from the nomadic life of hunter-gatherer bands to settled existence in much larger communities. This wrenching shift required living in a hierarchical society instead of an egalitarian one and the temperament to get on with many strangers instead of just a few close kin. Given that this change took so long to occur—modern humans first appear in the archaeological record 200,000 years ago, yet it took them 185,000 years to settle down in fixed communities—it is tempting to assume that a substantial genetic change in social behavior was required and that it took this long to evolve. Moreover, this evolutionary process took place independently in the populations of Europe, East Asia, the Americas and Africa, which had separated long before the first settlements emerged.

The forager-settler transition is unlikely to have been the only evolutionary change in human social behavior. Probably from the beginning of agriculture some 10,000 years ago, most people have lived on the edge of starvation. After each new increase in productivity, more babies were born, the extra mouths ate up the surplus and within a generation everyone was back to a state of scarcity little better than before.

This situation was accurately described by the Reverend Thomas Malthus with his analysis that population was always kept in check by misery and vice. It was from Malthus that Darwin derived the idea of natural selection. Under conditions of the fierce struggle for existence that Malthus described, favorable variations would be preserved, Darwin perceived, and unfavorable ones destroyed, leading eventually to the formation of new species.

Given that the human population supplied Malthus with the observations that led Darwin to the concept of natural selection, there is every reason to suppose that people living in agrarian societies were subject to intense forces of natural selection. But what traits were being selected for during the long agrarian past? Evidence described in chapter 7 indicates that it was human social nature that changed. Until the great demographic transition that followed industrialization, the wealthy had more surviving children than the poor. As many of the children of the rich fell in status, they would have spread throughout the population the genes that support the behaviors useful in accumulating wealth. This ratchet of wealth provides a general mechanism for making a specific set of behaviors—those required for economic success—more general and, generation after generation, gradually changing a society's nature. The mechanism has so far been documented only for a population for which unusually precise records exist, that of England from 1200 to 1800. But given the strong human propensity for investing in one's children's success, the ratchet may well have operated in all societies in which there have been gradations of wealth.

The narratives constructed by historians describe many forms of change, whether political, military, economic or social. One factor almost always assumed to be constant is human nature. Yet if human social nature, and therefore the nature of human societies, has changed in the recent past, a new variable is available to help explain major turning points in history. The Industrial Revolution, for instance, marked a profound change in the productivity of human societies, one that took almost 15,000 years to emerge after the first settlements. Could this too have required the evolution of a difference in human social behavior, as significant as the one that accompanied the transition from foraging to settled life?

There are other significant turning points in history for which scholars have proposed a clutch of possible causes but no compelling explanation. China created the first modern state and enjoyed the most advanced civilization until around 1800 AD, when it slid into puzzling decline. The Islamic world in 1500 AD surpassed the West in most respects, reaching a high tide of its expansion in the siege of Vienna in 1529 AD by the forces of the Ottoman Sultan Suleiman the Magnificent. Then, after almost a thousand years of relentless conquest, the house of Islam entered a long and painful retreat, also for reasons that defy scholarly consensus.

The counterpart of Chinese and Islamic decline is the unexpected rise of the West. Europe, feudal and semitribal in 1000 AD, had become a vigorous exponent of learning and exploration by 1500 AD. From this basis, Western nations seized the lead in geographical expansion, in military preeminence, in economic prosperity and in science and technology.

Economists and historians have described many factors that contributed to Europe's awakening. One that is seldom considered is the possibility of an evolutionary change, that the European population, in adapting to its particular local circumstances, happened to evolve a kind of society that was highly favorable to innovation.

## Economic Disparities

Explanation is also lacking for many important features of even today's world. Why are some countries rich and others persistently poor? Capital and information flow fairly freely, so what is it that prevents poor countries from taking out a loan, copying every Scandinavian institution, and becoming as rich and peaceful as Denmark? Africa has absorbed billions of dollars in aid over the past half century and yet, until a recent spurt of growth, its standard of living has stagnated for decades. South Korea and Taiwan, on the other hand, almost as poor at the start of the period, have enjoyed an economic resurgence. Why have these countries been able to modernize so rapidly while others have found it much harder?

Economists and historians attribute the major disparities between countries to factors such as resources or geography or cultural differences. But many countries with no resources, like Japan or Singapore, are very rich, while richly endowed countries like Nigeria tend to be quite poor. Iceland, covered mostly in glaciers and frigid deserts, might seem less favorably situated than Haiti, but Icelanders are wealthy and Haitians beset by persistent poverty and corruption. True, culture provides a compelling and sufficient explanation for many such differences. In the natural experiment provided by the two Koreas, the people are the same in both countries, so it must surely be bad institutions that keep North Koreans poor and good ones that make South Koreans prosperous.

But in situations where culture and political institutions can flow freely across borders, long enduring disparities are harder to explain. The brisk and continuing pace of human evolution suggests a new possibility: that at the root of each civilization is a particular set of evolved social behaviors that sustains it, and these behaviors are reflected in the society's institutions. Institutions are not just sets of arbitrary rules. Rather, they grow out of instinctual social behaviors, such as the propensity to trust others, to follow rules and punish those who don't, to engage in reciprocity and trade, or to take up arms against neighboring groups. Because these behaviors vary slightly from one society to the next as the result of evolutionary pressures, so too may the institutions that depend on them.

This would explain why it is so hard to transfer institutions from one society to another. American institutions cannot be successfully implanted in Iraq, for instance, because Iraqis have different social behaviors, including a base in tribalism and a well-founded distrust of central government, just as it would be impossible to import Iraqi tribal politics into the United States.

With the advent of fast and cheap methods for decoding the sequence of DNA units in the human genome, the genetic variations that underlie human races can be explored for the first time. The evolutionary paths that have generated differences between races are of great interest to researchers and many are described in the pages that follow. But the broader significance of the worldwide variations in DNA is not the differences but the similarities. Nowhere is the essential unity of humankind more clearly and indelibly written than in the human genome.

––––––

Since much of the material that follows may be new or unfamiliar to the general reader, a guide to its evidentiary status may be helpful. Chapters 4 and 5, which explore the genetics of race, are probably the most securely based. Although they put the reader on the forefront of current research, and frontier science is always more prone to upset than that in the textbooks, the findings reported here draw from a large body of research by leading experts in the field and seem unlikely to be revised in any serious way. Readers can probably take the facts in these chapters as reasonably solid and the interpretations as being in general well supported.

The discussion of the roots of human social behavior in chapter 3 also rests on substantial research, in this case mostly studies of human and animal behavior. But the genetic underpinnings of human social behavior are for the most part still unknown. There is therefore considerable room for disagreement as to exactly which social behaviors have a genetic basis and how strongly any such behaviors may be genetically defined. Moreover the whole field of research into human social behavior is both young and overshadowed by the paradigm still influential among social scientists that all human behavior is purely cultural.

Readers should be fully aware that in chapters 6 through 10 they are leaving the world of hard science and entering into a much more speculative arena at the interface of history, economics and human evolution. Because the existence of race has long been ignored or denied by many researchers, there is a dearth of factual information as to how race impinges on human society. The conclusions presented in these chapters fall far short of proof. However plausible (or otherwise) they may seem, many are speculative. There is nothing wrong with speculation, of course, as long as its premises are made clear. And speculation is the customary way to begin the exploration of uncharted territory because it stimulates a search for the evidence that will support or refute it.

# 2

◆━◆

# PERVERSIONS
# OF SCIENCE

Imperialists, calling upon Darwinism in defense of the subju-
gation of weaker races, could point to *The Origin of Species*,
which referred in its subtitle to *The Preservation of Favored
Races in the Struggle for Life*. Darwin had been talking about
pigeons, but the imperialists saw no reason why his theories
should not apply to men.

—RICHARD HOFSTADTER[1]

Ideas about race, many of them generated by biologists, have been
exploited to justify slavery, to sterilize people deemed unfit and,
in Hitler's Germany, to conduct murderous campaigns against
innocent and defenseless segments of society such as Gypsies, homosexuals and mentally ill children. Most chilling of all was the horrific
fusion of eugenic ideas with notions of racial purity that drove the
National Socialists to slaughter some 6 million Jews in the territories
under their control.

Because there could be no more serious caution for any who would inquire into the nature of race, the errors that lured people and governments down these mistaken paths need first to be understood.

Racism is a surprisingly modern concept, the word first appearing in the *Oxford English Dictionary* only in 1910. Before that, ethnic prejudice existed in profusion and still does. The ancient Greeks applied the word barbarian to anyone who didn't speak Greek. China has long called itself the Central Kingdom, regarding as barbarians all who live outside its borders. The click-speaking bushmen of the Kalahari Desert divide the world into Jul'hoansi, or "real people," such as themselves, and !ohm, a category that includes other Africans, Europeans and inedible animals such as predators. Europeans link nationality with edibility in devising derogatory names for one another. Thus the French refer to the English as *les rosbifs* ("roast beefs"), while the English call the French *frogs* (as in frogs' legs, a French delicacy) and Germans *krauts* (from sauerkraut, or fermented cabbage).

The central premise of racism, which distinguishes it from ethnic prejudice, is the notion of an ordered hierarchy of races in which some are superior to others. The superior race is assumed to enjoy the right to rule others because of its inherent qualities.

Besides superiority, racism also connotes the idea of immutability, thought once to reside in the blood and now in the genes. Racists are concerned about intermarriage ("the purity of the blood") lest it erode the basis of their race's superiority. Since quality is seen as biologically inherent, the racist's higher status can never be challenged, and inferior races can never redeem themselves. The notion of inherent superiority, which is generally absent from mere ethnic prejudice, is held to justify unlimited abuse of races held to be inferior, from social discrimination to annihilation. "The essence of racism is that it regards individuals as superior or inferior because they are imagined to share physical, mental and moral attributes with the group to which they are deemed to belong, and it is assumed that they cannot change these traits individually," writes the historian Benjamin Isaac.[2]

It's not surprising that the notion of racial superiority emerged in the 19th century, after European nations had established colonies in much of the world and sought a theoretical justification of their dominion over others.

At least two other strands of thought fed into modern ideologies of racism. One was the effort by scientists to classify the many human populations that European explorers were able to describe. The other was that of Social Darwinism and eugenics.

## Classifying Human Races

In the 18th century Linnaeus, the great classifier of the world's organisms, recognized four races, based principally on geography and skin color. He named them *Homo americanus* (Native Americans), *Homo europaeus* (Europeans), *Homo asiaticus* (East Asians) and *Homo afer* (Africans). Linnaeus did not perceive a hierarchy of races, and he listed people alongside the rest of nature.

In a 1795 treatise called *On the Natural Variety of Mankind*, the anthropologist Johann Blumenbach described five races based on skull type. He added a Malay race, essentially people of Malaya and Indonesia, to Linnaeus's four, and he invented the useful word Caucasian to denote the peoples of Europe, North Africa and the Indian subcontinent. The origin of the name was due in part to his belief that the people of Georgia, in the southern Caucasus, were the most beautiful and in part to the then prevailing view that Noah's ark had set down on Mount Ararat in the Caucasus, making the region the homeland of the first people to colonize the earth.

Blumenbach has been unjustly tarred with the supremacist beliefs of his successors. In fact he opposed the idea that some races were superior to others, and he conceded that his appraisal of Caucasian comeliness was subjective.[3] His views on Caucasian beauty can more reasonably be ascribed to ethnic prejudice than to racism. Moreover Blumenbach insisted that all humans belonged to the same species, as against the view then emerging that the human races were so different from one another as to constitute different species.

Up until Blumenbach, the study of human races was a reasonably scientific attempt to understand and explain human variation. The more dubious turn taken in the 19th century was exemplified by Joseph-Arthur Comte de Gobineau's book *An Essay on the Inequality of Human Races*, published in 1853–55. Gobineau was a French aristocrat and diplomat, not a scientist, and a friend and correspondent of de Tocqueville. His book was a philosophical attempt to explain the rise and fall of nations, based essentially on the idea of racial purity. He assumed there were three races recognized by the skin colors of white, yellow and black. A pure race might conquer its neighbors, but as it interbred with them, it would lose its edge and risk being conquered in turn. The reason, Gobineau supposed, was that interbreeding leads to degeneracy.

The superior race, Gobineau wrote, was that of the Indo-Europeans, or Aryans, and their continuance in the Greek, Roman and European empires. Contrary to what might be expected from Hitler's exploitation of his works, Gobineau greatly admired Jews, whom he described as "a people that succeeded in everything it undertook, a free, strong, and intelligent people, and one which, before it lost, sword in hand, the name of an independent nation, had given as many learned men to the world as it had merchants."

Gobineau's ambitious theory of history was built on sand. There is no factual basis for his notions of racial purity or racial degeneration through interbreeding. His ideas would doubtless have been forgotten but for the pernicious theme of an Aryan master race. Hitler adopted this worthless concept while ignoring Gobineau's considerably more defensible observations about Jews.

To Gobineau's assertion of inequality between races was then added the divisive idea that the various human populations represented not just different races but also different species. A leading proponent of this belief was the Philadelphia physician Samuel Morton.

Morton's views were driven into error not by prejudice but by his religious faith. He was troubled by the fact that black and white people were depicted in Egyptian art from 3000 BC yet the world itself had been created only in 4004 BC, according to the widely accepted chronology drawn up by Archbishop Ussher from information derived from the Old Testament and elsewhere. This was not enough time for different races to emerge, so the races must have been created separately, Morton argued, a valid inference if Ussher's chronology had been even remotely correct.

Morton amassed a large collection of skulls from all over the world, measuring the volume occupied by the brain and other details that in his view established the distinctness of the four principal races. He effectively ranked them in a hierarchy by adding subjective descriptions of each race's behavior to his careful anatomical measurements of their skulls. Europeans are the earth's "fairest inhabitants," he wrote. Next were Mongolians, meaning East Asians, deemed "ingenious, imitative and highly susceptible of cultivation." Third place was assigned to Americans, meaning Native Americans, whose mental faculties appeared to Morton as locked in a "continual childhood," and fourth were Negroes, or Africans, who Morton said "have little invention, but strong powers of imitation, so that they readily acquire mechanic arts."

Morton was an academic and did not promote any practical consequences of his ideas. But his followers had no hesitation in spelling out their interpretation that the races had been created separately, that blacks were inferior to whites and that the slavery of the American South was therefore justified.

Morton's data present an interesting case study of how a scientist's preconceptions can affect his results, despite the emphasis in scientific training on the critical importance of objectivity. The Harvard biologist Stephen Jay Gould, a widely read essayist, accused Morton of having mismeasured the cranial volumes of African and Caucasian skulls in order to support the view that brain size is related to intelligence. Gould didn't remeasure Morton's skulls, but he recomputed Morton's published statistical analysis and estimated that all four races had skull volumes of about the same size. Gould's accusations were published in *Science* and in his widely cited 1981 book *The Mismeasure of Man*.

But in a surprising recent twist, the bias now turns out to have been Gould's. Morton did not in fact believe, as Gould asserted, that intelligence was correlated with brain size. Rather, he was measuring his skulls to study human variation as part of his inquiry into whether God had created the human races separately. A team of physical anthropologists remeasured all of the skulls they could identify in Morton's collection and found his measurements were almost invariably correct. It was Gould's statistics that were in error, they reported, and the errors lay in the direction of supporting Gould's incorrect belief that there was no difference in cranial capacity between Morton's groups. "Ironically, Gould's own analysis of Morton is likely the stronger example of a bias influencing results," the Pennsylvania team wrote.[4]

The authors noted that "Morton, in the hands of Stephen Jay Gould, has served for 30 years as a textbook example of scientific misconduct." Moreover Gould had suggested that science as a whole is an imperfect process because bias such as Morton's is common. This, the authors suggested, is incorrect: "The Morton case, rather than illustrating the ubiquity of bias, instead shows the ability of science to escape the bounds and blinders of cultural contexts."

There are two lessons to be drawn from the Morton-Gould imbroglio. One is that scientists, despite their training to be objective observers, are as fallible as anyone else when their emotions or politics are involved, whether they come from the right or, as in Gould's case, from the left.

A second is that, despite the personal failings of some scientists, science as a knowledge-generating system does tend to correct itself, though often only after considerable delay. It is during these delay periods that great harm can be caused by those who use uncorrected scientific findings to propagate injurious policies. Scientists' attempts to classify human races and to understand the proper scope of eugenics were both hijacked before the two fields could be fully corrected.

A firm refutation of the idea that human races were different species was supplied by Darwin. In *On the Origin of Species,* published in 1859, he laid out his theory of evolution but, perhaps preferring to take one step at a time, said nothing in particular about the human species. Humans were covered in his second volume, *The Descent of Man,* which appeared 12 years later. With his unerring good sense and insight, Darwin decreed that the human races, however distinct they might appear, were not nearly different enough to be considered separate species, as the followers of Samuel Morton and others were contending.

He started out by observing that "if a naturalist, who had never before seen a Negro, Hottentot, Australian or Mongolian, were to compare them . . . he would assuredly declare that they were as good species as many to which he had been in the habit of affixing specific names."

In support of such a view (Darwin is making the best contrary case before he knocks it down), he noted that the various human races are fed on by different kinds of lice. "The surgeon of a whaling ship in the Pacific"—Darwin had a far-flung network of informants—"assured me that when the Pediculi, with which some Sandwich Islanders on board swarmed, strayed onto the bodies of the English sailors, they died in the course of three or four days." So if the parasites on human races are distinct species, it "might fairly be urged as an argument that the races themselves ought to be classified as distinct species," Darwin suggested.

On the other hand, whenever two human races occupy the same area, they interbreed, Darwin noted. Also, the distinctive traits of each race are highly variable. Darwin cited the example of the extended labia minora ("Hottentot apron") of bushmen women. Some women have the apron, but not all do.

The strongest argument against treating the races of men as separate species, in Darwin's view, "is that they graduate into each other, independently in many cases, as far as we can judge, of their having intercrossed." This graduation is so extensive that people trying to enumerate the number of human races were all over the map in their estimates, which ranged from 1 to 63, Darwin noted. But every naturalist trying to describe a group of highly varying organisms will do well to unite them into a single species, Darwin observed, for "he has no right to give names to objects which he cannot define."

Anyone reading works of anthropology can hardly fail to be impressed by the similarities between the races. Darwin noted "the pleasure which they all take in dancing, rude music, acting, painting, tattooing and otherwise decorating themselves; in their mutual comprehension of gesture-language, by the same expression in their features, and by the same inarticulate cries, when excited by the same emotions." When the principle of evolution is accepted, "as it surely will be before long," Darwin wrote hopefully, the dispute as to whether humans belong to a single species or many "will die a silent and unobserved death."

## Social Darwinism and Eugenics

Darwin, by force of his authority, could put the idea of many human species to rest. Despite his best efforts, he had less success in throttling the political movement called Social Darwinism. This was the proposition that just as in nature the fittest survive and the weak are pushed to the wall, the same rule should prevail in human societies too, lest nations be debilitated by the poor and sick having too many children.

The promoter of this idea was not Darwin but the English philosopher Herbert Spencer. Spencer developed a theory about the evolution of society, which held that ethical progress depended on people adapting to current conditions. The theory was developed independently of Darwin's and lacked any of the extensive biological research on which Darwin's was based. Still, it was Spencer who coined the phrase "survival of the fittest," which Darwin adopted.

Spencer argued that government aid that would allow the poor and sickly to propagate would impede society's adaptation. Even government support for education should be cut off, lest it postpone the elimination of those who failed to adapt. Spencer was one of the most prominent intellectuals of the second half of the 19th century, and his ideas, however harsh they may seem today, were widely discussed in both Europe and America.

Darwin's theory of evolution, at least in its author's eyes, dealt solely with the natural world. Yet it was as attractive to political theorists as a candle's flame is to moths. Karl Marx asked if he could dedicate *Das Kapital* to Darwin, an honor the great naturalist declined.[5] Darwin's name was slapped on to Spencer's political ideas, which would far more accurately have been called Social Spencerism. Darwin himself demolished them in a lapidary reproof.

Yes, vaccination has saved millions whose weaker constitutions would otherwise have let them succumb to smallpox, Darwin wrote. And yes, the weak members of civilized societies propagate their kind, which, to judge from animal breeding, "must be highly injurious to the race of man." But the aid we feel impelled to give to the helpless is part of our social instincts, Darwin said. "Nor could we check our sympathy, even at the urging of hard reason, without deterioration in the noblest part of our nature," he wrote. "If we were intentionally to neglect the weak and helpless, it could only be for a contingent benefit, with an overwhelming present evil."[6]

Had Darwin's advice been heeded, a disastrous turn in 20th century history might have been somewhat less inevitable. But for many intellectuals, theoretical benefits often outweigh overwhelming present evils. Airy notions of racial improvement drove the eugenics movement, which over many decades created the mental climate for the mass exterminations conducted by the National Socialists in Germany. Yet this catastrophe started out in such a different place. It started with Darwin's cousin, Francis Galton.

Galton was a Victorian gentleman and polymath who made distinguished contributions to many fields of science. He invented several basic statistical techniques, such as the concepts of correlation, regression and standard deviation. He anticipated human behavior genetics by using twins to sort out the influences of nature and nurture. He devised the classification scheme still used in fingerprint identification. He drew the first weather map. Mischievously wondering how to test if prayers were answered, he noted that the English population had for centuries prayed each week in church for the long life of their sovereign, so that if prayer had any power at all, it should surely result in the greater longevity of English monarchs. His report that sovereigns were the shortest-lived of all rich people and hence that prayer had no positive effect was rejected by an editor as "too terribly conclusive and offensive not to raise a hornet's nest" and lay unpublished for many years.[7]

One of Galton's principal interests was that of whether human abilities are hereditary. He compiled various lists of eminent people and looked for those who were related to one another. Within these families, he found that close relatives of the founder were more likely to be eminent than distant ones, establishing that intellectual distinction had a hereditary basis.

Galton was compelled by contemporary critics to pay more attention to the fact that the children of eminent men had greater educational and other opportunities than others. He conceded that nurture was involved to some extent, even inventing the phrase "nature versus nurture" in doing so. But his interest in the inheritance of outstanding abilities remained. Darwin's theory of evolution was now widely accepted in England and Galton, with his avidity for measuring human traits, was interested in the effect of natural selection on the English population.

This line of thought now led him down a dangerous path, to the proposal that human populations could be improved by controlled breeding, just like those of domestic animals. His finding that eminence ran in families led him to propose that marriages between such families should be encouraged with monetary incentives so as to improve the race. For this goal, Galton coined another word, eugenics.

In an unpublished novel, "Kantsaywhere," Galton wrote that those who failed eugenic tests were to be confined to camps where they had to work hard and remain celibate. But this seems to have been mostly a thought experiment or fantasy in Galton's mind. In his published work, he emphasized public education about eugenics and positive incentives for marriage among the eugenically fit.

There is no particular reason to doubt the assessment of one of Galton's biographers, Nicholas Gillham, that Galton "would have been horrified had he known that within little more than 20 years of his death forcible sterilization and murder would be carried out in the name of eugenics."[8]

Galton's ideas seemed reasonable at the time, given the knowledge of the day. Natural selection seemed to have loosened its grip on modern populations. Birth rates at the end of the 19th century were in decline, particularly sharply among the upper and middle classes. It seemed logical enough that the quality of the population would be improved if the upper classes could be encouraged to have more children. Galton's ideas were favorably received. Honors flowed in. He was awarded the Darwin Medal of the Royal Society, England's preeminent scientific institution. In 1908, three years before his death, he received a knighthood, a mark of establishment approval. No one then understood that he had unwittingly sown the dragon's teeth.

The lure of Galton's eugenics was his belief that society would be better off if the intellectually eminent could be encouraged to have more children. What scholar could disagree with that? More of a good thing must surely be better. In fact it is far from certain that this would be a desirable outcome. Intellectuals as a class are notoriously prone to fine-sounding theoretical schemes that lead to catastrophe, such as Social Darwinism, Marxism or indeed eugenics.

By analogy with animal breeding, people could no doubt be bred, if it were ethically acceptable, so as to enhance specific desired traits. But it is impossible to know what traits would benefit society as a whole. The eugenics program, however reasonable it might seem, was basically incoherent.

And in terms of practicalities, it held a fatal diversion. Galton's idea of eugenics was to induce the rich and middle class to change their marriage habits and bear more children. But positive eugenics, as such a proposal is known, was a political nonstarter. Negative eugenics, the segregation or sterilization of those deemed unfit, was much easier to put into practice.

In 1900 Mendel's laws of genetics, ignored in his lifetime, were rediscovered. Geneticists, by combining his laws with the statistical methods developed by Galton and others, started to develop the powerful subdiscipline known as population genetics. Leading geneticists on both sides of the Atlantic used their newfound authority to promote eugenic ideas. In doing so, they unleashed an idea whose deeply malignant powers they proved unable to control.

The principal organizer of the new eugenics movement was Charles Davenport. He earned a doctorate in biology from Harvard and taught zoology at Harvard, the University of Chicago, and the Brooklyn Institute of Arts and Sciences Biological Laboratory at Cold Spring Harbor on Long Island. Davenport's views on eugenics were motivated by disdain for races other than his own: "Can we build a wall high enough around this country so as to keep out these cheaper races, or will it be a feeble dam . . . leaving it to our descendants to abandon the country to the blacks, browns and yellows and seek an asylum in New Zealand?" he wrote.[9]

A heavy wave of immigrants arrived in the United States between 1890 and 1920, creating a climate of concern that was favorable for eugenic ideas. Davenport, though he had no special distinction as a scientist, found it easy to raise money for his eugenics program. He secured funds from leading philanthropies, such as the Rockefeller Foundation and the recently founded Carnegie Institution. Scouring a list of wealthy families on Long Island, he came across the name of Mary Harriman, daughter of the railroad magnate E. H. Harriman. Mary, as it happened, was so interested in eugenics that her nickname in college had been Eugenia. She provided Davenport funds to set up his Eugenics Record Office, which was intended to register the genetic backgrounds of the American population and distinguish between good and defective lineages.[10]

The Carnegie and Rockefeller institutions don't give money to just anyone, but rather to fields of research that their advisers judge promising. These advisers shared the generally favorable view of eugenics that then prevailed among scientists and many intellectuals. The Eugenics Research Association included members from Harvard, Columbia, Yale and Johns Hopkins.[11]

"In America, the eugenic priesthood included much of the early leadership responsible for the extension of Mendelism," writes the science historian Daniel Kevles. "Besides Davenport, there were Raymond Pearl and Herbert S. Jennings, both of Johns Hopkins University; Clarence Little, the president of the University of Michigan and later the founder of the Jackson Laboratory in Maine; and the Harvard professors Edward M. East and William E. Castle. . . . The large majority of American colleges and universities—including Harvard, Columbia, Cornell, Brown, Wisconsin, Northwestern, and Berkeley—offered well-attended courses in eugenics, or genetics courses that incorporated eugenic material."[12]

The same conclusion is drawn by another historian of the eugenics movement, Edwin Black: "The elite thinkers of American medicine, science and higher education were busy expanding the body of eugenic knowledge and evangelizing its tenets," he wrote.[13]

Where so many eminent scientists led, others followed. Former president Theodore Roosevelt wrote to Davenport in 1913, "We have no business to permit the perpetuation of citizens of the wrong type."[14] The eugenics program reached a pinnacle of acceptance when it received the imprimatur of the U.S. Supreme Court. The court was considering an appeal by Carrie Buck, a woman whom the State of Virginia wished to sterilize on the grounds that she, her mother and her daughter were mentally impaired.

In the 1927 case, known as *Buck v. Bell*, the Supreme Court found for the state, with only one dissent. Justice Oliver Wendell Holmes, writing for the majority, endorsed without reservation the eugenicists' credo that the offspring of the mentally impaired were a menace to society.

"It is better for the world," he wrote, "if instead of waiting to execute degenerate offspring for crime, or to let them starve for their imbecility, society can prevent those who are manifestly unfit from continuing their kind. The principle that sustains compulsory vaccination is broad enough to cover cutting the Fallopian tubes. Three generations of imbeciles are enough."

Eugenics, having started out as a politically impractical proposal for encouraging matches among the well-bred, had now become an accepted political movement with grim consequences for the poor and defenseless.

The first of these were sterilization programs. At the urging of Davenport and his disciples, state legislatures passed programs for sterilizing the inmates of their prisons and mental asylums. A common criterion for sterilization was feeblemindedness, an ill-defined diagnostic category that was often identified by knowledge-based questions that put the ill educated at particular disadvantage.

Eugenicists perverted intelligence tests into a tool for degrading people. The tests had been first developed by Alfred Binet to recognize children in need of special educational help. The eugenics movement used them to designate people as feebleminded and hence fit for sterilization. Many of the early tests probed knowledge, not native wit. Questions like "The Knight engine is used in the: Packard/Stearns/ Lozier/Pierce Arrow" or "Becky Sharp appears in: Vanity Fair/ Romola/A Christmas Carol/Henry IV" were heavily loaded against those who had not received a particular kind of education. As Kevles writes, "The tests were biased in favor of scholastic skills, and the outcome was dependent upon the educational and cultural background of the person tested."[15] Yet tests like these were used to destroy people's hopes of having children or deny them entry into military service.

Up until 1928, fewer than 9,000 people had been sterilized in the United States, even though the eugenicists estimated that up to 400,000 citizens were "feeble minded."[16] After the *Buck v. Bell* decision, the floodgates opened. By 1930, 24 states had sterilization laws on their books, and by 1940, 35,878 Americans had been sterilized or castrated.[17]

Eugenicists also began to influence the nation's immigration laws. The 1924 Immigration Act pegged each country's quota to the proportion of its nationals present in the 1890 census, a reference point later changed to the 1920 census. The intent and effect of the law was to increase immigration from Nordic countries and restrict people from southern and eastern Europe, including Jews fleeing persecution in Poland and Russia. In addition, the act barred all immigration from most East Asian countries. As Congressman Robert Allen of West Virginia explained during the floor debate, "The primary reason for restriction of the alien stream . . . is the necessity for purifying and keeping pure the blood of America."[18]

The eugenicists had inspectors installed in the major capitals of Europe to screen prospective immigrants. Almost a tenth were judged to be physically or mentally defective. The inspectorate collapsed after a few years because of its expense, but its preferences lingered on in the minds of U.S. consuls. When Jews in increasing numbers tried to flee Germany after 1936, U.S. consuls refused to grant visas to them and other desperate refugees.[19]

Many supporters of the 1924 Immigration Act were influenced by a book called *The Passing of the Great Race*. Its author, Madison Grant, was a New York lawyer and conservationist who helped found the Save the Redwoods League, the Bronx Zoo, Glacier National Park and Denali National Park. Despite his lack of scholarly credentials, Grant was powerful in anthropological circles and clashed frequently with Franz Boas, the founder of American social anthropology and a champion of the idea that significant differences between societies are cultural, not biological, in origin. Grant tried to get Boas fired from his position as chair of the anthropology department at Columbia University and fought a losing campaign with him over control of the American Anthropological Association.

Grant's beliefs were starkly racist and eugenic. He considered that Europeans, based on the skull and other physical traits, consisted of three races, which he called Nordic, Alpine and Mediterranean. The Nordics, with their brown or blond hair and blue or pale eyes, were the superior type, in part because the harsh northern climate in which they evolved "must have been such as to impose a rigid elimination of defectives through the agency of hard winters and the necessity of industry and foresight in providing the year's food, clothing and shelter during the short summer."

It followed that "such demands on energy if long continued would produce a strong, virile and self-contained race which would inevitably overwhelm in battle nations whose weaker elements had not been purged."[20]

England's decline was due to the "lowering proportion of its Nordic blood and the transfer of political power from the vigorous Nordic aristocracy and middle classes to the radical and labor elements, both largely recruited from the Mediterranean type," Grant wrote. The "master race" was threatened by the same dilution in the United States: "Apparently America is doomed to receive in these later days the least desirable classes and types from each European nation now exporting men."

Emma Lazarus saw the United States as a beacon of hope for the refugees from Europe's savage wars and hatreds. Grant had a less expansive vision to offer: "We Americans must realize that the altruistic ideals which have controlled our social development during the past century and the maudlin sentimentalism that has made America 'an asylum for the oppressed,' are sweeping the nation toward a racial abyss. If the Melting Pot is allowed to boil without control and we continue to follow our national motto and deliberately blind our-selves to all 'distinctions of race, creed or color,' the type of native American of Colonial descent will become as extinct as the Athenian of the age of Pericles, and the Viking of the days of Rollo."[21]

Grant's book was little read by the 1930s, when Americans began to turn against eugenic ideas. But its shaping of the 1924 Immigra-tion Act was not the least of its malignant effects. Grant received a fan letter one day from an ardent admirer who had incorporated many ideas from *The Passing of the Great Race* into a work of his own. "The book is my Bible," the writer assured Grant. Grant's fan, the author of *Mein Kampf,* was Adolf Hitler.[22]

The drift toward eugenics was not inexorable. In England, eugenic ideas never left the realm of theory. The Galtonian version of eugenics at first attracted a wide following among the intelligentsia, including the playwright George Bernard Shaw and social radicals such as Beatrice and Sidney Webb. Winston Churchill, then home secretary, told eugenicists during discussion of the Mental Deficiency Act of 1913 that Britain's 120,000 citizens deemed feebleminded "should, if possible, be segregated under proper conditions so that their curse died with them and was not transmitted to future genera-tions."

But Parliament did not favor sterilization. In 1931 and 1932 the Eugenics Society managed to get bills introduced to allow voluntary sterilization, but they went nowhere. There was no taste for such extreme measures and, in any case, surgical sterilization of anyone, even with the person's consent or that of a court-appointed guardian, would have been considered a criminal act under English law.

The Eugenics Society in Britain had far less success in influencing public opinion than Davenport's eugenic lobby did in the United States. One reason was that most English scientists, after an initial infatuation with Galton's ideas, turned against eugenics, particularly the kind being promoted by Davenport.

Davenport believed that ill-defined traits such as "shiftlessness" or "feeblemindedness" were caused by single genes and had the simple patterns of inheritance that Mendel had described in his experimental pea plants. But complex behavioral traits are generally governed by many genes acting in concert. While a Mendelian trait could in principle be almost eliminated by sterilizing its carriers, were it ethical to do so, complex traits are much harder to influence in this way.

A 1913 article by a member of the Galton laboratory, David Heron, attacked certain American work for "careless presentation of data, inaccurate methods of analysis, irresponsible expression of conclusions and rapid change of opinion." Many recent contributions to the subject, in the writer's view, threatened to place eugenics "entirely outside the pale of true science."[23]

The English critics were correct about the quality of Davenport's science, although it continued to hold sway in the United States for many years more. When the Carnegie Institution got around to obtaining an objective review of Davenport's work at the Eugenics Record Office in 1929, its reviewers too found that the office's data were worthless. A second review committee concluded in 1935 that eugenics was not a science and that the Eugenics Record Office "should devote its entire energies to pure research divorced from all forms of propaganda and the urging or sponsoring of programs for social reform or race betterment such as sterilization, birth control, inculcation of race or national consciousness, restriction of immigration, etc."

By 1933, eugenics had reached a fateful turning point. In both England and the United States, scientists had first embraced the idea and then turned against it, followed by their respective publics. Eugenics might have withered to a mere footnote in history if scientists in Germany had followed their colleagues in rejecting eugenic ideas. Hitler's rise to power foreclosed any such possibility.

German eugenicists kept in close touch with their American colleagues both before and after the First World War. They saw that American eugenicists favored Nordic races and sought to keep the gene pool unsullied. They watched with keen interest as many state legislatures in the United States set up programs for sterilizing the mentally disabled, and as Congress changed the immigration laws to favor immigrants from northern Europe over other regions of the world.

U.S. eugenic laws and ideology "became inspirational blueprints for Germany's rising tide of race biologists and race-based hatemongers," wrote the author Edwin Black.[24] Hitler came to power on January 30, 1933, and Germany's eugenics program quickly got under way. In the Law for the Prevention of Defective Progeny, decreed on July 14, 1933, Germany identified nine categories of people to be sterilized—the feebleminded and those with schizophrenia, manic depression, Huntington's disease, epilepsy, deafness, hereditary deformities, hereditary blindness and alcoholism. The latter aside, these were the same illnesses targeted by Davenport and the American eugenicists.

Some 205 local Hereditary Health Courts were set up in Germany, each with three members—a lawyer who served as chairman, a eugenicist and a physician. Doctors who failed to report suspect patients were fined. Sterilizations began on January 1, 1934, and covered children over ten and people at large, not just those in institutions. During the first year, 56,000 people were sterilized. By 1937, the last year that records were published, the total had reached 200,000 people.

The purpose of the 1933 law, according to an official at the Reich Ministry of the Interior, was to prevent "poisoning the entire bloodstream of the race." Sterilization would safeguard the purity of the blood in perpetuity. "We go beyond neighborly love; we extend it to future generations," the official said. "Therein lies the high ethical value and justification of the law."[25]

The sterilization program involved doctors and hospitals and created a legal and medical system for coercive treatment of those whom the National Socialists deemed unfit. With this machinery in place, it was much easier to extend the eugenics program in two major directions. One was the transition from sterilization to killing, prompted in part by the growing shortage of hospital beds as the Second World War got under way. In 1939 some 70,000 mentally disabled patients in asylums were designated for euthanasia. The first victims were shot. Later ones were forced into rooms disguised as showers, where they were gassed.[26]

The other departure in Germany's eugenics program was the addition of Jews to the list of those considered unfit. A succession of punitive laws drove Jews from their jobs and homes, isolated them from the rest of the population, and then confined those who had not already fled to concentration camps where they were murdered.

The first anti-Jewish decree, of April 7, 1933, provided for the dismissal of "non-Aryan" civil servants. The term "non-Aryan" offended foreign nations such as Japan. Future laws referred to Jews explicitly but plunged the Reich Ministry of the Interior into the problem of deciding who was a Jew. The National Socialist Party proposed that half-Jews be considered Jews, but the Ministry of the Interior rejected the idea as impractical. It divided half-Jews into two categories, considering them as full Jews only if they belonged to the Jewish religion or were married to a Jew. Using this definition, the Nuremberg Law of September 13, 1935, otherwise known as the Law for the Protection of German Blood and Honor, prohibited marriage between Jews and citizens of "German or related blood."[27]

These measures were followed by others that in a few years escalated to a program of mass murder of Jews in Germany and the European countries occupied by Hitler's troops. Of the 9 million Jews who lived in Europe before the Holocaust, nearly 6 million were killed, including 1 million children. The killing machine engulfed a further

4 to 5 million victims in the form of homosexuals, Gypsies and Russian prisoners of war. It was Hitler's aim to depopulate the countries of Eastern Europe so as to make room for German settlers.

Many of the elements in the National Socialists' eugenics program could be found in the American eugenics program, at least in concept, though not in degree. Nordic supremacy, purity of the blood, condemnation of intermarriage, sterilization of the unfit—all these were ideas embraced by American eugenicists.

The destruction of the Jews, however, was Hitler's idea. So too was the replacement of sterilization with mass murder.

The fact that antecedents for the ideas that led to the Holocaust can be found in the American and English eugenics movements of the 1920s and 1930s does not mean that others share responsibility for the crimes of the National Socialist regime. It does mean that ideas about race are dangerous when linked to political agendas. It puts responsibility on scientists to test rigorously the scientific ideas that are placed before the public.

In Germany, scientists played a major role in paving the way for the destruction of the Jews but were not solely culpable. Anti-Semitic statements mar the writings of leading German philosophers, including even Kant. Wagner ranted against the Jews in his operas and essays. "By the end of the First World War," writes Yvonne Sherratt in her survey of intellectual influences on Hitler, "anti-Semitic ideas pervaded every aspect of German thought from the Enlightenment to Romanticism, from nationalism to science. Men of logic or the passions, Idealists or Social Darwinists, the highly sophisticated or the very crude, all supplied Hitler with the ideas to re-inforce and enact his dream."[28] Anti-Semitism was not an idea that German scientists found in science; rather, they found it in their culture and allowed it to infect their science.

*Scientia* means "knowledge," and true scientists are those who distinguish meticulously between what they know scientifically and what they don't know or may only suspect. Those involved with Davenport's eugenics program, including his sponsors at the Carnegie Institution and the Rockefeller Foundation and their reviewers, failed to say immediately that Davenport's ideas were scientifically defective. Scientists' silence or inattention allowed a climate of public opinion to develop in which Congress could pass restrictive immigration laws, state legislatures could decree the sterilization of those judged mentally infirm and the U.S. Supreme Court could uphold unwarranted assaults on the country's weakest citizens.

After the Second World War, scientists resolved for the best of reasons that genetics research would never again be allowed to fuel the racial fantasies of murderous despots. Now that new information about human races has been developed, the lessons of the past should not be forgotten and indeed are all the more relevant.

# 3

---

# ORIGINS OF HUMAN
# SOCIAL NATURE

Humankind's behavioral unity exists, but it lies deeply buried
under several thousand years of cumulative cultural evolution
and is barely visible from the human realm.

—BERNARD CHAPAIS[1]

It deserves notice that, as soon as the progenitors of man
became social . . . the principles of imitation, and reason, and
experience would have increased, and much modified the
instinctual powers in a way, of which we see only traces in
the lower animals.

—CHARLES DARWIN[2]

One of the strangest features of human anatomy, when people
are compared with the other 200 monkey and ape species in
the primate family, is the sclera, or the white of the eye. In all
our primate cousins, the sclera is barely visible. In humans it stands
out like a beacon, signaling to any observer the direction of a person's gaze and hence what thoughts may be on their mind.

Why should such a feature have evolved? A signal that reveals a person's thoughts to a competitor or to an enemy on the battlefield can be a deadly handicap. For natural selection to have favored it, there must be a compensating advantage of overwhelming magnitude. And that advantage must have something to do with the social nature of the interaction, the abundant benefit conferred on all members of a group by being able to infer what others are thinking just by sizing up the direction of their gaze. The whites of the eyes are the mark of a highly social, highly cooperative species whose success depends on the sharing of thoughts and intentions.

Human sociality is often assumed to be entirely a matter of culture, originating from the age of life when children are taught to be nice to one another. A cascade of discoveries, many in the past decade, has made clear that this is not the case. Human sociality has been shaped by natural selection, just as might be expected for any feature so crucial to survival. Sociality is written into our physical form, as with the whites of the eyes and the self-mortifying phenomenon of blushing as a signal of embarrassment. It is engraved in our neural circuitry too, most obviously in the faculty of language—there is no point in talking to oneself—and in many other behaviors. These include an inclination to follow rules and an urge to punish others when they fail to do so. Shame and guilt are the penalties for our own failings. To achieve status and avert retribution, we are always seeking to burnish our reputation. We trust the members of our in-group and are prepared to distrust the out-group. We often know instinctively what is right and wrong.

The genes that set up the circuitry of these social instincts have not yet been identified, but their presence can be inferred from several lines of evidence that are described below. The salient fact is that all types of human society, from the hunter-gatherer band to the modern nation, are rooted in a suite of social behaviors. These behaviors, which most probably have a genetic basis, interact with culture to produce the institutions that are characteristic of each society and help it survive in its particular environment.

Any trait that has a genetic basis can be changed by natural selection. The existence of genes that have some bearing on human social behavior means that social behavior can be reworked by evolution and therefore can vary in time and place. But natural selection's remodeling of human societies is far harder to identify than changes in skin color, say, because skin color depends primarily on the genes whereas social behavior, harder to measure in any case, is strongly influenced by culture.

Nonetheless, it is reasonable to assume that if traits like skin color have evolved in a population, the same may be true of its social behavior, and hence the very different kinds of society seen in the various races and in the world's great civilizations differ not just because of their received culture—in other words, in what is learned from birth—but also because of variations in the social behavior of their members, carried down in their genes.

Given the vast power of culture to shape human social behavior, it's necessary to look far back in evolutionary history to glimpse the signs of social behavior genes at work.

# From Chimpanzee Society to Human Society

The nature of human society can most clearly be understood by tracing how it evolved. Humans and chimpanzees, our closest evolutionary cousins, split apart some 5 to 6 million years ago. There is reason to think that the joint ancestor of humans and chimps was far more chimplike than human. Chimpanzees seem to live in much the same habitat as they did 5 million years ago, and their basic way of life hasn't changed. The apes at the head of the human lineage, on the other hand, abandoned the forest and ventured out into the open savannahs of Africa, obliging them to go through many evolutionary transitions in both body and behavior as they grew more and more unlike the joint ancestor they shared with chimps.

If the joint ancestor of chimps and humans was chimplike, so too was its social behavior. The society of living chimps can thus with reasonable accuracy stand in as a surrogate for the society of the joint ancestor and hence describe the baseline from which human social behavior evolved.

Chimp bands are hierarchical. An alpha male and one or two allies dominate the male hierarchy, and below that is a less visible female hierarchy. The males are fiercely territorial, probably to protect the fruit trees that are the community's chief source of food. Females usually stay and feed in one region of the territory. The larger each female's region is and the more fruit trees it contains, the more children she can bear.

To maintain and increase the size of their territory, male chimps conduct regular patrols around its perimeter, with occasional forays into their neighbors' territory. Male chimps are unremittingly hostile to strange males and if possible will kill them on sight. Their favorite tactic on invading enemy territory is to surprise and kill any male whom they find alone. If the raiding party senses that it is outnumbered, it will retreat. A neighboring territory will be captured after its resident males have been killed off one by one in a campaign that may last several years.

Chimp reproductive behavior requires a female to mate with all the males in her band, or at least as many as possible. She is estimated to copulate between 400 and 3,000 times per conception. This labor provides an insurance policy for her children, since each male who thinks he might be the father of her child is more likely to refrain from killing it.

Despite the flamboyant promiscuity of female chimps, the alpha male somehow manages to fulfill his droit du seigneur of fathering many of the community's offspring—about 36% in one study based on DNA paternity tests, or 45% excluding the close female relatives with whom he would avoid mating. The high-ranking males who were his allies together scored 50% of paternities.

An important feature of chimp communities is that the females mostly disperse to neighboring groups when they reach adolescence, while the males stay in the community where they were born, an arrangement called patrilocality. Dispersal at puberty, which serves to avoid inbreeding, is common in primate communities, except that most are matrilocal, meaning that it's the males who disperse and the females who stay in their home community. Chimps, many hunter-gatherer societies, and to some extent gorillas, are patrilocal. This arrangement probably has much to do with the chimp and human propensity for warfare: a group of males who have grown up together will be more cohesive in defending their own territory against rival groups. Since the males need to stay together, this obliges the females to move so as to avoid inbreeding.

A strange feature of chimp society, at least from the human perspective, is that kinship is almost invisible. If you are born into a chimp society, you will know your mother and the siblings born a few years before or after you, because these are the chimps who hang out around your mother. But you will have no idea who your father is, though he must be one of the males in the community, nor any notion of who his relatives are, even though you see them every day. You are equally ignorant of your mother's relatives, whom she left behind in her home community when she migrated to yours as a teenager. When a chimp raiding party enters neighboring territory, the males it kills may often be relatives or in-laws of the invaders' daughters and sisters who dispersed there. But this kinship is unknown to the raiders.

How then was the profound transition made from the chimplike society of the joint ancestor to the hunter-gatherer societies in which all humans lived until 15,000 years ago and in which kinship was a central institution? The likely steps in this process have been persuasively worked out by the primatologist Bernard Chapais. The critical behavioral step, in his view, was formation of the pair bond, or at least a stable breeding relationship between male and female.

Consider a population of chimplike creatures living in a forest in Africa more than 5 million years ago. A fierce drought gripped Africa from 6.5 to 5 million years ago, and the forests shrank, giving way to open woodland or savannah. This was perhaps the event that forced the population apart into two groups, one of which led to chimps and the other to humans. In response to the drought, some of the population clung to the traditional habitat and became the ancestors of chimps. Others left the trees and sought new sources of food on the ground, despite the risk of being caught in the open by large cats and other predators. This group became the ancestors of the human lineage.

The group trying life on the ground eventually started to walk upright, probably because walking on two feet is more efficient than knuckle-walking, the great ape method of making the knuckles of the hands serve as a pair of forefeet. Freeing the hands, though an accidental by-product of walking upright, was an adaptation of far-reaching significance because the hands could now be used for gripping tools and for gesturing.

Another adaptation, equally accidental and far-reaching, led to a transformation of social structure. This was the practice of mate guarding, which developed into the formation of stable breeding relationships and eventually of the pair bond between one male and one female.

Males of almost all primate species, even chimps, guard females to some extent, so as to deter other males and improve their own chances of fathering the females' children. Among the population of chimplike ancestors that had left the trees, mate guarding would have become more common than usual because of the more dangerous environment on the ground.

With the male often around for defense, he could also help in feeding and taking care of the children. Having at least two people involved in child rearing made an enormous difference, Chapais argues. The period of juvenile dependence could last for several years longer. Children could be born at an earlier stage in their development since they would be more protected, and earlier birth enabled the brain to do more of its growing outside the womb. The human brain eventually reached three times the size of that of chimps.

At first each male guarded as many females as he could, but another development drove them unwillingly toward monogamy. This was the emergence of weapons. At first, physical strength was decisive in fending off other males. But weapons are great equalizers because they tend to negate the advantages of size. The cost of maintaining a large harem became too high for most males. Weapons forced most to settle for one wife. The pair bond between male and female became established.

Having a dad around makes all the difference to social networks. In highly promiscuous societies like those of chimps, an individual knows only its mother and the siblings it grows up with. With pair bonding, people know not only their father as well as their mother, but all their father's relatives too. The males in a community now recognized both their daughters and, when their daughters dispersed to a neighboring group, a daughter's husband and his parents.

The neighbors, who used to be treated as hostile, began now to be seen in an entirely different light. Those males, who once had to be killed on sight, were not the enemy—they were the in-laws, with an equal interest in promoting the welfare of one's daughter's or sister's children. Thus in the incipient human line, a new and more complex social structure came into being, that of the tribe, a group of bands bound to one another by exchange of women.

Warfare between neighboring bands, the chimp practice, was now pushed upward to the tribal level. Tribes would fight as savagely as before, but among the bands within each tribe cooperation was now the rule.

This profound transition in social structure started some time after the split of the ancestral populations leading to chimps and humans. Pair bonding, an essential element of the new social structure, probably did not become significant until the emergence of *Homo ergaster* some 1.7 million years ago. This is the first human ancestor in which the males were not very much larger than the females. A large size difference between the sexes, as in gorillas, indicates competition between males and a harem structure. The size difference diminishes as pair bonding becomes more common.

Given the distinctiveness of chimp social behavior, there is no reason to doubt that it has a genetic basis. Both the chimp and human lineages would have inherited a suite of genes governing social behavior and in each species the genes for social behavior would have evolved as social structure changed in response to the society's requirements for survival.

Chimp social structure, in fact, may not differ much from that of the joint chimp-human ancestor. But human social structure has changed profoundly over the past 5 million years. Just as physical form was changing from ape to human, social behavior was undergoing a radical transformation from the chimplike behavior of multimale bands to the human pair-bond system. There is every reason to suppose that the development of distinctive social behavior in humans had a genetic basis, just as surely as the physical changes did. And if social behavior was under genetic control during the evolution of human society from that of a chimplike ancestor, it is hard to see why it should not have continued to be molded by evolutionary forces up until the present day.

Social behavior changes in response to changes in the environment. As the hominid groups abandoned the trees, for eons the primates' safe refuge, their societies had to adapt to the richer opportunities and more serious perils of life on the ground. This highly risky endeavor required a thorough makeover of standard ape social behavior, most pertinently in the degree of cooperation between individuals.

# The Distinctive Human Virtue: Cooperation

Chimps will cooperate in certain ways, like assembling in war parties to patrol the borders of their territory. But beyond the minimum requirements for being a social species, they have little instinct to help one another. Chimps in the wild forage for themselves. Even chimp mothers regularly decline to share food with their children, who are able from a young age to gather their own food. When the mothers do share food, it's always the rind or husk or less desirable part that is given to the child.[3]

In the laboratory, chimps don't naturally share food either. With some exceptions, most experiments show chimps to be severely lacking in altruistic sentiments for other chimps. If a chimp is put in a cage where he can pull in one tray of food for himself or, with no greater effort, a tray that also provides food for a neighbor in the next cage, he will pull indiscriminately—he just doesn't care whether his neighbor gets fed or not. Yet he's perfectly aware that one of the trays carries a portion of food made available to the neighboring cage. If the cage next door is empty and the chimp is allowed access to it, he will usually pull the tray with the double portion. Chimps are truly selfish.[4]

Human children, on the other hand, are inherently cooperative. From the earliest ages, they desire to help others, to share information and to participate in pursuing common goals. The developmental psychologist Michael Tomasello has studied this cooperativeness in a series of experiments with very young children. He finds that if infants aged 18 months see an unrelated adult with hands full trying to open a door, almost all will immediately try to help. If the adult pretends to have lost an object, children from as young as 12 months will helpfully point out where it is.

There are several reasons to believe that the urges to help, inform and share are "naturally emerging" in young children, Tomasello writes, meaning that they are innate, not taught.[5] One is that these instincts appear at a very young age before most parents have started to train their children to behave socially. Another is that the helping behaviors are not enhanced if the children are rewarded.

A third reason is that social intelligence develops in children before their general cognitive skills, at least when compared with apes. Tomasello gave human and chimp children a battery of tests related to understanding the physical and social worlds. The human children, aged 2.5 years, did no better than the chimps on the physical world tests but were considerably better at understanding the social world.[6]

The essence of what children's minds have and chimps' don't is what Tomasello calls shared intentionality. Part of this ability is that they can infer what others know or are thinking, a skill called theory of mind. But beyond that, even very young children want to be part of a shared purpose. They actively seek to be part of a "we," a group that has pooled its talents and intends to work toward a shared goal.

Children of course have the selfish motivations necessary for survival, like any other animal, but a vigorous social instinct is overlaid on their behavior from a very young age. The social instinct gets modulated in later life as the children learn to make distinctions about whom they can trust and who does not reciprocate.

Besides shared intentionality, another striking social behavior is that of following norms, or rules generally agreed on within the "we" group. Allied with the rule following are two other basic principles of human social behavior. One is a tendency to criticize, and if necessary punish, those who do not follow the agreed-upon norms. Another is to bolster one's own reputation, presenting oneself as an unselfish and valuable follower of the group's norms, an exercise that may involve finding fault with others.

The first two behaviors are already evident in very young children. Tomasello showed a group of two- and three-year-olds a new game. A puppet then appeared and performed the game incorrectly. Almost all the children protested the puppet's actions and many explicitly objected, telling the puppet how the game should be played. "Social norms—even of this relatively trivial type—can only be created by creatures who engage in shared intentionality and collective beliefs," Tomasello writes, "and they play an enormously important role in maintaining the shared values of human cultural groups."[7]

The urge to punish deviations from social norms is a distinctive feature of human societies. In principle it carries great risks for the punisher. In tribal or hunter-gatherer societies, anyone who punishes a miscreant is likely to have vengeance wreaked upon him by the miscreant's family. So punishment in practice is meted out fairly deliberately. First, through social gossip, a consensus is arrived at that an individual's behavior merits correction. Punishment may then be carried out collectively, by shunning or even ostracizing the deviant member. A different problem arises when the offender refuses to reform and must be killed. Hunter-gatherers will usually persuade his family to do the job because anyone else will bring down a blood feud on his head.

Social norms and punishment of deviants are behaviors embedded so deeply in the human psyche that special mechanisms have arisen for punishing oneself for infractions of social norms: shame and guilt, which are sometimes physically expressed by blushing.

A delicate balance was being maintained during the evolution of human social structure. As human brain size increased, individuals could calculate to an ever greater degree where their own self-interest lay and how it might be served at the group's expense. To deter freeloading, ever more sophisticated countermechanisms were required. Together with shame and guilt, an inbuilt sense of morality evolved, one that gave people an instinctive aversion to murder and other crimes, at least against members of their own group. A propensity for religious behavior bound people together in emotion-laden rituals that affirmed commitment to common goals. And religion instituted a vigilant overseer of people's actions, a divine avenger who would punish infractions with disaster in this world and torment in the next.

As these mechanisms for group cohesion evolved, humans became the most social of animals, and their societies, growing ever more capable, began the series of achievements that was to lead eventually to the first settlements and agriculture.

## The Hormone of Social Trust

Human sociability is two-edged. The trust extended to members of one's own group is mirrored by the suspicion and potential mistrust shown toward strangers. Willingness to defend one's own people is the counterpart of readiness to kill the enemy. Human morality is not universal, as philosophers have argued: it is strictly local, at least in its instinctual form. Reflections of this ambivalence are now apparent from the level of the genes.

If human social nature is innate and has evolved, as seems highly likely, there will be evidence of its evolution in the genome. Very little about the genes that govern the human brain is yet understood, so it need be no surprise that not much is yet known about the genetic basis of human social behavior. A prominent exception concerns the neural hormone called oxytocin, sometimes known as the hormone of trust. It is synthesized in a region in the base of the brain known as the hypothalamus and from there is distributed to both brain and body, with separate roles in each. In the body, oxytocin is released when a woman gives birth and when she gives milk to her child.

In the brain, oxytocin has a range of subtle effects that are only beginning to be explored. In general, it seems that oxytocin has been co-opted in the course of evolution to play a central role in social cohesion. It's a hormone of affiliation. It dampens down the distrust usually felt toward strangers and promotes feelings of solidarity. "It increases men's trust, generosity and willingness to cooperate," say the authors of a recent review.[8] (The same is doubtless true of women too but most such experiments are performed only in men because of the risk that oxytocin might make a woman miscarry if she were unknowingly pregnant.)

The trust promoted by oxytocin is not of the brotherhood of man variety—it's strictly local. Oxytocin engenders trust toward members of the in-group, together with feelings of defensiveness toward outsiders. This limitation in oxytocin's radius of trust was discovered only recently by Carsten De Dreu, a Dutch psychologist who doubted the conventional wisdom that oxytocin simply promoted general feelings of trust. Any individual who blindly trusts everyone is not going to prosper in the struggle for survival, De Dreu supposed, and his genes would be rapidly eliminated; hence it seemed much more likely that oxytocin promoted trust only in certain contexts.

De Dreu showed in several ingenious experiments that this is indeed the case. In one, the young Dutch men who were his subjects were presented with standard moral dilemmas, such as whether to save five people in the path of a train for the loss of one life, that of a bystander who could be thrown onto the tracks to stop the train. The people to be saved were all Dutch but the person to be killed was sometimes given a Dutch first name, like Pieter, and sometimes a German or Muslim name, like Helmut or Muhammad. (Opinion polls show that neither is a favorite nationality among the Dutch.)

When the subjects had taken a sniff of oxytocin, they were much more inclined to sacrifice the Helmuts and the Muhammads, De Dreu found, showing the dark side of oxytocin in making people more willing to punish outsiders. Oxytocin does not seem to promote positive aggression toward outsiders, he finds, but rather it heightens the willingness to defend the in-group.[9]

The two-edged nature of oxytocin is just what might be expected to suit the needs of ancestral humans living in small tribal groups where every stranger was a possible enemy. In larger societies, for instance in cities, where people must often do business with strangers, the general level of trust needs to be considerably higher than in tribal societies, where most interactions are with close kin.

So deep are oxytocin's roots that it is involved in the most basic aspect of human sociality, that of recognizing people's faces. Doses of oxytocin improve a subject's recognition of human faces. Genetic variations in the gene that specifies the oxytocin receptor protein are associated with impaired face recognition.[10]

When oxytocin reaches a target neuron, it interacts with a receptor protein that juts out from the neuron's surface and is crafted to recognize oxytocin specifically. The strength with which these receptors bind to oxytocin can be varied by making small changes in the receptor's gene. An experiment to test this of course cannot be done in people, but relevant evidence comes from comparing two species of vole. Male prairie voles are monogamous and make caring, trustworthy fathers, whereas male meadow voles are roaming polygamists who leave much to be desired in the fatherhood realm. If meadow voles are genetically engineered so as to stud their neurons with extra receptors for vasopressin, a hormone very similar to oxytocin, these Lotharios suddenly become monogamous.[11]

It's easy to see how natural selection could increase the general degree of trust in human societies, whether by raising the brain's production of oxytocin, by inserting more oxytocin receptors into people's neurons, or by enhancing the tenacity with which the receptors hold on to oxytocin. The opposite processes would lower the degree of social trust. It is not yet known by what specific mechanism the oxytocin levels in people are controlled. But the oxytocin mechanism can evidently be modulated by natural selection so as to achieve either more or less of the same effect. If an inclination to distrust others should favor survival, people with lower oxytocin levels will flourish and have more children, and in several generations, a society will become less trusting. Conversely, if stronger bonds of trust help a society flourish, genes that increase oxytocin levels will become more common.

This is not to imply that trust in human societies is set exclusively by the genes. Culture is far more important in most short-term interactions. As with most human behaviors, the genes provide just a nudge in a certain direction. But these small nudges, acting on every individual, can alter the nature of a society. Small changes in social behavior can, in the long term, deeply modify the social fabric and make one society differ significantly from another.

## Control of Aggression

Besides trust, another important social behavior that is clearly under genetic influence is that of aggression, or rather the whole spectrum of behaviors that runs from aggression to shyness. The fact that animals can be domesticated is proof that the trait can be modulated by the selective pressures of evolution.

One of the most dramatic experiments on the genetic control of aggression was performed by the Soviet scientist Dmitriy Belyaev. From the same population of Siberian gray rats he developed two strains, one highly sociable and the other brimming with aggression. For the tame rats, the parents of each generation were chosen simply by the criterion of how well they tolerated human presence. For the ferocious rats, the criterion was how adversely they reacted to people. After many generations of breeding, the first strain was now so tame that when visitors entered the room where the rats were caged, the animals would press their snouts through the bars to be petted. The other strain could not have been more different. The rats would hurl themselves screaming toward the intruder, thudding ferociously against the bars of their cage.[12]

Rodents and humans use many of the same genes and brain regions to control aggression. Experiments with mice have shown that a large number of genes are involved in the trait, and the same is certainly true of people. Comparisons of identical twins raised together and separately show that aggression is heritable. Genes account for between 37% and 72% of the heritability, the variation of the trait in a population, according to various studies. But very few of the genes that underlie aggression have yet been identified, in part because when many genes control a behavior, each has so small an effect that it is hard to detect. Most research has focused on genes that promote aggression rather than those at the other end of the behavioral spectrum.

One of the genes associated with aggression is called MAO-A, meaning that it makes one of two forms of an enzyme called monoamine oxidase. The enzyme has a central role in maintaining normal mental states through its cleanup function—it breaks down three of the small neurotransmitter chemicals used to convey signals from one neuron to another. The three neurotransmitters, serotonin, norepinephrine, and dopamine, need to be disposed of after accomplishing their signaling task. If allowed to accumulate in the brain, they will keep neurons activated that should have returned to rest.

The role of MAO-A in the control of aggression came to light in 1993 through the study of a Dutch family in which the men were inclined to violently deviant behavior, such as impulsive aggression, arson, attempted rape and exhibitionism. The eight affected men had inherited an unusual form of the MAO-A gene. A single mutation in the gene causes the cell's assembly of the MAO-A enzyme to be halted halfway through, rendering it ineffective. In the absence of functioning MAO-A enzymes, neurotransmitters build up in excess, causing the men to be overaggressive in social situations.[13]

Mutations that totally disrupt a gene like MAO-A have serious consequences for the individual. There are more subtle ways in which a gene like MAO-A can be modulated by natural selection so as to make people either more or less aggressive. Genes are controlled by elements called promoters, which are short stretches of DNA that lie near the genes they control. And being made of DNA, the promoters can incur mutations just like the DNA of the genes.

As it happens, the promoter for MAO-A is quite variable in the human population. People may have two, three, four or five copies of it, and the more copies they have, the more of the MAO-A enzyme their cells produce. What difference does this make to a person's behavior? Quite a lot, it turns out. People with three, four or five copies of the MAO-A promoter are normal but those with only two copies have a much higher level of delinquency. From a questionnaire given to 2,524 youths in the United States, Jean Shih and colleagues found that men with just two promoters were significantly more likely to report that they had committed both serious delinquency within the previous 12 months, such as theft, selling drugs or damaging property, and violent assaults, such as hurting someone badly enough to need medical care or threatening someone with a knife or gun.

Women with two promoters also reported much higher levels of serious and violent delinquency than those with more promoters.[14]

If individuals can differ in the genetic structure of their MAO-A gene and its controls, is the same also true of races and ethnicities? The answer is yes. A team led by Karl Skorecki of the Rambam Health Care Center in Haifa looked at variations in the MAO-A gene in people from seven ethnicities—Ashkenazi Jews, Bedouins, African pygmies, aboriginal Taiwanese, East Asians (Chinese and Japanese), Mexicans and Russians. They found 41 variations in the portions of the gene they decoded, and the pattern of variation differed from one ethnicity to the next, revealing a "substantial differentiation between populations."

The pattern of variation could have arisen from random mutations in the DNA that had no effect on the MAO-A enzyme or on people's behavior. But after applying various tests, the researchers concluded there was possible evidence for "positive selection, potentially acting on MAO-A-related phenotypes."[15] This means they think that natural selection could have favored particular behavioral traits in the various ethnicities, whether more or less aggressive, and that this could have caused the particular patterns of variation in the MAO-A gene. But the researchers did not examine the behaviors of the various ethnicities so could not establish causal links between each pattern of variations in the MAO-A enzyme and specific behavioral traits.

Such a link has been asserted by a research team led by Michael Vaughn of Saint Louis University. He and his colleagues looked at the MAO-A promoters in African Americans. The subjects were the same 2,524 American youths in the study by Shih mentioned above. Of the African American men in the sample, 5% carried two MAO-A promoters, the condition that Shih had found to be associated with higher levels of delinquency. Members of the two-promoter group were significantly more likely to have been arrested and imprisoned than African Americans who carried three or four promoters. The same comparison could not be made in white, or Caucasian, males, the researchers report, because only 0.1% carry the two-promoter allele.[16]

A finding like this has to be interpreted with care. First, like any scientific report, it needs to be repeated by an independent laboratory to be sure it is valid. Second, a large number of genes are evidently involved in controlling aggression, so even if African Americans are more likely to carry the violence-linked allele of MAO-A promoters than are Caucasians, Caucasians may carry the aggressive allele of other genes yet to be identified. Indeed a variant of a gene called HTR2B, an allele that predisposes carriers to impulsive and violent crimes when under the influence of alcohol, has been found in Finns.[17] It is therefore impossible, by looking at single genes, to say on genetic grounds that one race is genetically more prone to violence than any other. Third, genes don't determine human behavior; they merely create a propensity to behave in a certain way. Whether a propensity to violence is exercised depends on circumstances as well as genetic endowment, so that people who live in conditions of poverty and unemployment may have more inducements to violence than those who are better off.

The wider point illustrated by the case of the MAO-A gene is that important aspects of human social behavior are shaped by the genes and that these behavior traits are likely to vary from one race to another, sometimes significantly so.


# How Societies Change to Fit Environment

Trust and aggression are two significant components of human social behavior whose underlying genetics have already been to some extent explored. There are many other aspects of social behavior, such as conformity to rules, the willingness to punish violators of social norms or the expectation of fairness and reciprocity, that most probably have a genetic basis, although one that remains to be discovered.

The fact that human social behavior is to some extent shaped by the genes means that it can evolve and that different kinds of society can emerge as the underlying social behaviors shift. Conversely, major changes in human society, such as the transition from hunter-gathering to settled life, were almost certainly accompanied by evolutionary changes in social behavior as people adapted to their new way of life. (The words adapt and adaptation are always used here in the biological sense of a genetically based evolutionary response to circumstances.)

There are two important factors to consider in the emergence of social change. One is that a society develops through changes in its institutions, which are blends of culture and genetically shaped social behavior. The other is that the genes and culture interact. This may seem paradoxical to anyone who considers genes and culture to be entirely separate realms. But it is scarcely surprising from an evolutionary perspective, given that the genome is designed to respond to the environment, and a major component of the human environment is society and its cultural practices.

The working components of a society are its institutions. Any socially agreed-upon form of behavior, from a tribal dance to a parliament, may be considered an institution. Institutions reflect both culture and history, but their basic building blocks are human behaviors. Follow an institution all the way down, and beneath thick layers of culture, it is built on instinctual human behaviors. The rule of law would not exist if people didn't have innate tendencies to follow norms and to punish violators. Soldiers could not be made to follow orders were not army discipline able to invoke innate behaviors of conformity, obedience and willingness to kill for one's own group.

So consider the intricate dynamics of the natural system in which the members of a human society are embedded. Their basic motivation is their own survival and that of their families. Unlike species that can only interact directly with their environment, people often do so through their society and its institutions. In responding to an environmental change, a society adjusts its institutions, and its members adjust to the new institutions by changing their culture in the short term and their social behavior in the long term.

The idea that human behavior has a genetic basis has long been resisted by those who see the mind as a blank slate on which only culture can write. The blank slate notion has been particularly attractive to Marxists, who wish government to mold socialist man in its desired image and who see genetics as an impediment to the power of the state. Marxist academics led the attack on Edward O. Wilson when he proposed in his 1975 book *Sociobiology* that social behaviors such as conformity and morality had a genetic basis. Wilson even suggested that genes might have some influence "in the behavioral qualities that underlie variations between cultures."[18] Although his term sociobiology is not now widely used—evolutionary psychology is a less controversial term for much the same thing—the tide has turned in favor of Wilson's ideas now that many human faculties seem to be innate. From the social repertoire of babies to the moral instincts discernible from psychological tests, it is clear that the human mind is hereditarily predisposed to act in certain ways.

Social behavior changes because, over a period of generations, genes and culture interact. "The genes hold culture on a leash," Wilson writes. "The leash is very long but inevitably values will be constrained in accordance with their effects on the human gene pool."[19] Harmful cultural practices may lead to extinction, but advantageous ones create selective pressures that can promote specific genetic variants. If a cultural practice provides a significant survival advantage, genes that enable a person to engage in that practice will become more common.

This interaction between the genome and society, known as

gene-culture evolution, has probably been a powerful force in shaping human societies. At present it has been documented for only minor dietary changes, but these establish the principle. The leading example is that of lactose tolerance, the ability to digest milk in adulthood by means of the enzyme lactase, which breaks down lactose, the principal sugar in milk.

In most human populations, the lactase gene is permanently switched off after weaning so as to save the energy required to make the lactase enzyme. Lactose, the sugar metabolized by the lactase

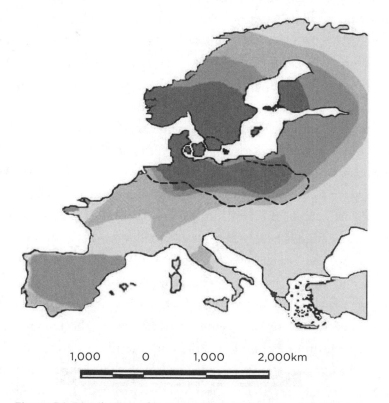

Figure 3.1. Distribution of lactose tolerance in present-day Europe (dark gray = 100%). Dotted area shows homeland of Funnel Beaker Culture, which flourished 6,000 to 5,000 years ago.

Source: Albano Beja-Pereira, *Nature Genetics* 35 (2003), pp. 311–15

enzyme, occurs only in milk, so that when a person has finished breast-feeding, lactase will never be needed again. But in populations that learned to herd cattle and drink raw cow's milk, notably the Funnel Beaker Culture that flourished in north central Europe from 6,000 to 5,000 years ago, there was a great selective advantage in keeping the lactase gene switched on. Almost all Dutch and Swedish people today are lactose tolerant, meaning they carry the mutation that keeps the lactase gene permanently switched on. The mutation is progressively less common in Europe with increasing distance from the core region of the ancient Funnel Beaker Culture.

Three different mutations that have the same result have been detected in pastoral peoples of eastern Africa. Natural selection has to work on whatever mutations are available in a population, and evidently different mutations were available in the European and various African peoples who took up cattle raising and drinking raw milk. The lactase-prolonging mutations conferred an enormous advantage on their owners, letting them leave ten times more surviving children than those without the mutation.[20]

Lactose tolerance is a fascinating example of how a human cultural practice, in this case cattle raising and drinking raw milk, can feed back into the human genome. The genes that underlie social behavior have for the most part not yet been identified, but it's a reasonable assumption that they too would have changed in response to new social institutions. In larger societies requiring a higher degree of trust, people who trusted only their close kin would have been at a disadvantage. People who were more trusting would have had more surviving children, and any genetic variation that promoted this behavior would become more common in each successive generation.

# The Shaping of Human Social Behavior

Changes in the social behaviors that underlie a society's institutions take many generations to accomplish. It may have been in hunting or scavenging that early humans first faced strong selective pressure to cooperate. Hunting is much more efficient when done as a group; indeed that's the only way that large game can be taken down, butchered and guarded from rivals. Hunting may have induced the shared intentionality that is characteristic of humans; groups that failed to cooperate closely did not survive. Along with cooperativeness emerged the rules for sharing meat in an equitable way and the gossip machinery that punished bragging and stinginess.

A hunter-gatherer society consists of small, egalitarian bands without leaders or headmen. This was the standard human social structure until 15,000 years ago. That it took 185,000 years for people to take the seemingly obvious step of settling down and putting a permanent roof over their heads strongly suggests that several genetic changes in social behavior had to evolve first. The aggressive and independent nature of hunter-gatherers, accustomed to trusting only their close kin, had to yield to a more sociable temperament and the ability to interact peaceably with larger numbers of people. A foraging society that turns to agriculture must develop a whole new set of institutions to coordinate people in the unaccustomed labor of sowing and harvesting crops.

In this novel environment, people skilled in farming and in operating in larger communities prospered and left more children; those whose only skill was in hunting did less well and placed fewer of their children and genes in the next generation. In time, the nature of the society and its members changed as its institutions were transformed to serve the new way of life.

After the first settlements, a wave of new societies then came into being in response to population pressure and new ways of gathering food. The anthropologist Hillard Kaplan and colleagues have worked out the dynamics of several of these adaptations.[21]

One reason why hunter-gatherer societies are egalitarian is that their usual food sources—game animals, tubers, fruits and nuts— tend to be dispersed and are not easily monopolized. In tribal horticulture, as practiced in New Guinea and parts of South America, people live in settled villages with gardens that must be planted and defended. This mode of life requires more structure than a hunter-gatherer band. People accept the governance of a headman to organize defense and conduct diplomatic relations with neighboring groups.

Tribal pastoralism creates an even greater demand for military leadership because the tribe's chief resource, herds of cattle or sheep, can easily be captured and driven off. Competition for grassland is another source of friction. Pastoralists have developed the necessary institutions for frequent warfare, which often include the social segregation of young warrior classes and expansionary male lineages.

The rise of the first city-states, based on large scale agriculture, required a new kind of social structure, one based on large, hierarchically organized populations ruled by military leaders. The states overlaid their own institutions on those of the tribe. They used religion to legitimate the ruler's power and maintain a monopoly of force.

The common theme of all these developments is that when circumstances change, when a new resource can be exploited or a new enemy appears on the border, a society will change its institutions in response. Thus it's easy to see the dynamics of how human social change takes place and why such a variety of human social structures exists. As soon as the mode of subsistence changes, a society will develop new institutions to exploit its environment more effectively. The individuals whose social behavior is better attuned to such institutions will prosper and leave more children, and the genetic variations that underlie such a behavior will become more common. If the pace of warfare increases, a special set of institutions will emerge so as to increase the society's military preparedness. These new institutions will feed back into the genome over the course of generations, as those with the social behaviors that are successful in a militaristic society leave more surviving children.

This process of continuous adaptation has taken a different course in each region of the world because each differed in its environment and exploitable resources. As population increased, coordinating the activities of larger numbers of people required more complex social structures. Tribes merged into archaic states, states became empires, and empires rose and fell, leaving behind the large scale structures known as civilizations.

The process of organizing people in larger and larger social structures, with accompanying changes in social behavior, has most probably been molded by evolution, though the underlying genetic changes have yet to be identified. This social evolution has proceeded roughly in parallel in the world's principal populations or races, those of Africans, East Asians and Caucasians. (Caucasian includes Europeans, the peoples of the Indian subcontinent and Middle Easterners.) The same process is visible in a fourth race, the natives of North and South America. Because the Americas were populated much later than Africa and Eurasia—the first settlers crossed the Bering Strait from Siberia only 15,000 years ago—social evolution got a much later start and the great empires of the Incas and Mayans emerged several thousand years later than their counterparts in Eurasia. In a fifth race, the peoples of Australia and Papua New Guinea, population numbers were always too low to ignite the processes of settlement and state building.

# How Evolution Creates Different Societies

People are entirely different from ants, yet there is something to be learned from the creatures that occupy the other pinnacle of social evolution in nature. An ant is an ant is an ant, yet natural selection has crafted a profusion of widely different ant societies, each adapted to its own ecological niche. Leaf-cutter ants are superb agricultural-ists, tending underground gardens of a mushroomlike fungus which they protect with special antibiotics. There are ants that live in the hollow thorns provided for them by acacia trees. Some ants specialize in preying on termite nests. Weaver ants sew leaves together to con-struct shelters for their colonies. Army ants kill every living thing that cannot escape from their intense raiding parties.

In the case of ants, evolution has generated their many different kinds of society by keeping the ant body much the same and altering principally the behavior of each society's members. People too live in many different types of society, and evolution seems to have con-structed these with the same strategy—keep the human body much the same but change the social behavior.

A principal difference is that people, with their far greater intelli-gence, construct societies full of complex interactions in which an individual with stereotyped behavior like an ant's would be at a severe disadvantage. Learned behavior, or culture, plays a dominant role in human societies, shaped by a small, though critical, set of geneti-cally influenced social behaviors. In ant societies, by contrast, social behavior is dominated by the genes and the genetically prescribed pheromones that govern the major activities of an ant society.

In human societies, individuals' behavior is therefore flexible and generalist, with much of a society's specificity being embedded in its culture. Human societies are not nearly as diverse as those of ants because evolution has had a mere 50,000 years in which to shape modern human populations, compared with the 100 million years of ant evolution.

Another major difference is that among people, individuals can generally move easily from one society to another. Ants will kill ants from other species or even a neighboring colony of the same species. Apart from slavery—some species of ant will enslave other species—ant societies are immiscible. The institutions of ant societies are shaped almost entirely by genetics and little, if at all, by culture. There is no way that army ants can be trained to stop raiding and turn to peaceful horticulture like leaf-cutter ants. With human societies, institutions are largely cultural and based on a much smaller genetic component.

In the case of both ants and people, societies evolve over time as natural selection modifies the social behavior of their members. With ants, evolution has had time to generate thousands of different species, each with a society adapted to survival in its particular environment. With people, who have only recently dispersed from their ancestral homeland, evolution has so far generated only races within a single species, but with several major forms of society, each a response to different environments and historical circumstances. New evidence from the human genome now makes it possible for the first time to examine this differentiation of the human population at the genetic level.

# 4

---•◆•---

# THE HUMAN
# EXPERIMENT

There is, however, no doubt that the various races, when care-
fully compared and measured, differ much from each other. . . .
The races differ also in constitution, in acclimatisation and in
liability to certain diseases. Their mental characteristics are
likewise very distinct; chiefly as it would appear in their emotional, but partly in their intellectual faculties.

—CHARLES DARWIN[1]

Through independent but largely parallel evolution among the
populations of each continent, the human species has differen-
tiated into races. This evolutionary process is hard to explore,
however, when the question of race is placed under taboo or its existence is denied outright.

Many scholars like to make safe nods to multicultural orthodoxy
by implying that human races do not exist. *Race? Debunking a Scientific Myth* is the title of a recent book by a physical anthropologist

and a geneticist, though their text is not nearly so specific.[2] "The concept of race has no genetic or scientific basis," writes Craig Venter, who was the leading decoder of the human genome but has no known expertise in the relevant discipline of population genetics.[3]

Only people capable of thinking the Earth is flat believe in the existence of human races, according to the geographer Jared Diamond. "The reality of human races is another commonsense 'truth' destined to follow the flat Earth into oblivion," he asserts.[4] For a subtler position, consider the following statement, which seems to say the same thing. "It is increasingly clear that there is no scientific basis for defining precise ethnic or racial boundaries," writes Francis Collins, director of the National Human Genome Research Institute in a review of the project's implications.[5] This form of words, commonly used by biologists to imply that they accept the orthodox political take on the nonexistence of race, means rather less than meets the eye. When a distinct boundary develops between races, they are no longer races but separate species. So to say there are no precise boundaries between races is like saying there are no square circles.

A few biologists have begun to agree that there are human races, but they hasten to add that the fact means very little. Races exist, but the implications are "not much," says the evolutionary biologist Jerry Coyne.[6] Too bad—nature has performed this grand 50,000 year experiment, generating scores of fascinating variations on the human theme, only to have evolutionary biologists express disappointment at her efforts.

From biologists' obfuscations on the subject of race, sociologists have incorrectly inferred that there is no biological basis for race, confirming their preference for regarding race as just a social construct. How did the academic world contrive to reach a position on race so far removed from reality and commonsense observation?

The politically driven distortion of scientific views about race can be traced to a sustained campaign from the 1950s onward by the anthropologist Ashley Montagu, who sought to make the word race taboo, at least when referring to people. Montagu, who was Jewish, grew up in the East End district of London, where he experienced considerable anti-Semitism. He was trained as a social anthropologist in London and New York, where he studied under Franz Boas, a champion of racial equality and the belief that culture alone shapes human behavior. He began to promote Boas's ideas with more zeal than their author. Montagu developed passionate views on the evils of race. "Race is the witchcraft, the demonology of our time, the means by which we exorcise the imagined demoniacal powers among us," he wrote. "It is the contemporary myth, humankind's most dangerous myth, America's Original Sin."[7]

In the postwar years, with the horror of the Holocaust weighing on people's minds, Montagu found ready acceptance of his views. These were prominent in the influential UNESCO statement on race, first issued in 1950, which he helped draft. He believed that imperialism, racism and anti-Semitism were driven by notions of race and could be undermined by showing that races did not exist. However much one may sympathize with Montagu's motives, it is perhaps simplistic to believe that an evil can be eliminated by banning the words that conceptualize it. But suppression of the word was Montagu's goal, and to a remarkable extent he succeeded.

"The very word race is itself racist," he wrote in his book *Man's Most Dangerous Myth: The Fallacy of Race*.[8] Many scholars who understood human races very well began to drop the use of the term rather than risk being ostracized as racists. In a survey taken in 1987, only 50% of physical anthropologists (researchers who deal with human bones) agreed that human races exist, and among social anthropologists (who deal with people) just 29% did so.

The physical anthropologists best acquainted with race are those who do forensics. Human skulls fall into three distinctive shapes, which reflect their owners' degree of ancestry in the three main races, Caucasian, East Asian and African. African skulls have rounder nose and eye cavities, and jaws that protrude forward, whereas Caucasians and East Asians have flatter faces. Caucasian skulls are longer, have larger chinbones and tear-shaped nose openings. East Asian skulls tend to be short and broad with wide cheekbones. There are many other features characteristic of the three skull types. As is often the case, there is no single feature that suffices to assign a skull to a particular racial type; rather, each feature is more common in one race than the others, allowing a combination of such features to be diagnostic.

By taking just a few measurements, physical anthropologists can tell police departments the race of a skull's former owner with better than 80% accuracy. This ability has occasioned some anguish among those persuaded by Montagu that human races shouldn't be acknowledged. How could they identify a skull's race so accurately if race doesn't exist? "That forensic anthropologists place our field's stamp of approval on the traditional and unscientific concept of race each time we make such a judgement is a problem for which I see no easy solution," wrote one physical anthropologist. His suggestion was to obfuscate, by retaining the concept but substituting a euphemism for the word race, such as ancestry.[9] This advice has been followed by a wide range of researchers who, while retaining the necessary concept of race, refer to it in print with bland periphrases like "population structure" or "population stratification." As for the actual DNA elements now used by biologists to assign people to their race, or races if of mixed parentage, these are known discreetly as AIMs, or ancestry informative markers.

# Evolution and Speciation

Races are a way station on the path through which evolution generates new species. The environment keeps changing, and organisms will perish unless they adapt. In the course of adaptation, different variations of a species will emerge in conditions where the species faces different challenges. These variations, or races, are fluid, not fixed. If the selective pressure that brought them into being should disappear, they will merge back into the general gene pool. Or, if a race should cease to interbreed with its neighbors through the emergence of some barrier to reproduction, it may eventually become a separate species.

People have not been granted an exemption from this process. If human differentiation were to continue at the same pace as that of the past 50,000 years, one or more of today's races might in the distant future develop into a different species. But the forces of differentiation seem now to have reversed course due to increased migration, travel and intermarriage.

Races develop within a species and easily merge back into it. All human races, so far as is known, have the same set of genes. But each gene comes in a set of different flavors or alternative forms, known to geneticists as alleles. One might suppose that races differ in having different alleles of various genes. But, though a handful of such racially defining alleles do exist, the basis of race rests largely on something even slighter, a difference in the relative commonness, or frequency, of alleles, a situation discussed further in the next chapter.

The frequency of each allele of a gene changes from one generation to the next, depending on the chance of which parent's allele is inherited and whether the allele is favored by natural selection. Races are therefore quite dynamic, because the allele frequencies on which they depend are shifting all the time. A good description is provided by the historian Winthrop Jordan in his history of the historical origins of racism in the United States. "It is now clear," he writes, "that mankind is a single biological species; that races are neither discrete nor stable units but rather that they are plastic, changing, integral parts of a whole that is itself changing. It is clear, furthermore, that races are best studied as products of a process; and, finally, that racial differences involve the relative frequency of genes and characteristics rather than absolute and mutually exclusive distinctions."[10]

Races emerge as part of the process of evolutionary change. At the level of the genome, the driving force of evolution is mutation. Mutation generates novelty in the sequence of DNA units that comprise the hereditary information. The new sequences are then acted on—either eliminated, made more common or ignored—by the evolutionary processes of natural selection, genetic drift and migration.

The chemical units of which DNA is composed are long lasting but not permanent. Every so often, from spontaneous decay or radiation, a unit will disintegrate. In every living cell, repair enzymes constantly patrol up and down the strands of DNA, proofreading the sequence of chemical units, or bases, as chemists call them. The four bases are known for short as A (adenine), T (thymine), G (guanine) and C (cytosine). The structure of a DNA molecule consists of two strands that spiral around each other in a double helix, with each base on one strand lightly cross-linked to a base on the other strand. The cross-linking system requires that where one strand of the double helix has A, there will be a T at the same site on the opposite strand, with G and C being similarly paired. If the base opposite a T is missing, the repair enzymes know to insert an A. If a C is missing its partner, the enzymes will provide a G. The system, though amazingly efficient, is not perfect. A wrong base is occasionally inserted by the proofreading system, and these "typos" are called mutations. When the mutations happen to occur in a person's germ cells, whether eggs or sperm, they become evolutionarily significant, because they may then get passed on to the next generation.

Other kinds of mutation occur through copying errors made by the cell in manipulating DNA. All these types of mutation are the raw material for natural selection, the second evolutionary force. Most mutations affect only the copious regions of DNA that lie between the genes and are of little consequence. It's the sequence of bases in the genes that codes the information that specifies proteins and other working parts of the cell. This coding DNA, as it is called, occupies less than 2% of the human genome. Mutations that do not meaningfully alter the coding DNA or the nearby promoter regions of DNA, used to activate the coding DNA, generally have no effect on the organism. Natural selection has no reason to bother about them, and for this reason geneticists call them neutral mutations.

Of the mutations that do change the genetic sequence, most degrade or even destroy the function of the protein specified by the gene. These mutations are detrimental and need to be eliminated. "Purifying selection" is the phrase geneticists use for the action of natural selection in ridding the genome of harmful mutations. The bearer of the mutation fails to live or has few or no offspring.

It's just a handful of mutations that have a beneficial effect, and these become more common in the population with each succeeding generation as the lucky owners are better able to survive and breed.

The individuals with a beneficial mutation possess a new gene, or rather a new allele—a version of the old gene with the new mutation embedded in it. It's because of mutations and alleles that there exists a third force of evolutionary change, called genetic drift. Each generation is a genetic lottery. Your father and mother each have two copies of every gene. Each parent bequeaths one of their two copies to you. The other is left on the cutting room floor. Suppose that with a particular gene there are just two versions, called alleles A and B, in a population. Suppose too that 60% of a present population carries allele A and 40% allele B. In the next generation, these proportions will change because, by the luck of the draw, allele A will be passed on to children more often than allele B, or the other way round.

If you follow the fate of allele A down the generations, it does a random walk in terms of its frequency in the population, from 60% in one generation, say, to 67% in the next to 58% to 33% and so on. But the walk cannot continue forever, because sooner or later it will hit one of two numbers, either 0% or 100%. If the frequency falls to 0%, allele A is permanently lost from the population. If it hits 100%, it's allele B that is lost and allele A that becomes the permanent form of the gene, at least until a new and better mutation crops up. This fluctuation in frequency is a random process known as genetic drift, and when the walk ends in allele A hitting 100%, geneticists say that it has become fixed or has gone to fixation, meaning it's the only game in town.

An important part of the genome that has gone to fixation is the DNA of the energy-producing mitochondria, former bacteria that were captured and enslaved long ago by the ancestor of all animal and plant cells. The mitochondria, little organelles within every cell, are inherited through the egg and passed down from a mother to her children. At some early stage in modern human evolution, one woman's mitochondrial DNA went to fixation by edging out all other versions of mitochondrial DNA.

The same winner-take-all victory was attained by a particular version of the Y chromosome, which men alone carry because it includes the male-determining gene. At a time when the human population was quite small, a single individual's Y chromosome increased in frequency until it became the only one left. As described below, the genetic legacies of the mitochondrial Eve and the Y-chromosomal

Adam have proved immensely useful for tracing the migration of their descendants around the globe.

This rise and fall of the alleles depends on the blind chance of which are cast aside and which pass into the next generation when the egg and sperm cells are created. Genetic drift can be a powerful force in shaping populations, particularly small ones, in which the drift toward either loss or fixation can happen within a few generations.

Another force that shapes the genetic heritage of a species is migration. As long as a population stays together and interbreeds, everyone draws from a common gene pool in which each gene exists in many different versions or alleles. An individual, however, can carry at most two alleles of each gene, one from each parent. So if a group of individuals breaks off from the main population, it will carry away only some of the alleles in the general pool, thus losing part of the available genetic endowment.

Mutation, drift, migration and natural selection are all unceasing forces that drive the engine of evolution ever onward. Even if a population stays in the same place and its phenotype, or physical form, remains the same, its genotype, or hereditary information, will remain in constant flux, running like the Red Queen to stay in the same place.

A population can stay more constant if it interbreeds, with everyone drawing from the same pool of alleles. As soon as any barrier to interbreeding occurs, such as a river encountered as the species spreads out, the populations on either side of the river will become subtly different from each other because of genetic drift. They will have taken the first step toward becoming subspecies, or races, and will continue to accumulate minor differences. Eventually one of these minor differences, perhaps a shift in the season of mating or in mate preference, will create a reproductive barrier between the two subspecies. As soon as individuals in the two populations cease to mate freely, the two subspecies are ready to split into distinct species.

# The Peopling of the World

So consider how this mechanism of differentiation, of a species developing into races, would have applied to humans. The change agents of migration, drift and natural selection bore down on the human population with particular force as soon as people started to disperse from the ancestral homeland. Those leaving Africa seem to have comprised a few hundred people, consisting perhaps of a single hunter-gatherer band. They took with them only a fraction of the alleles in the ancestral human population, making them less genetically diverse. They spread across the world by a process of population budding. When a group grew too big for the local resources, it would split, with one band staying put and the other moving a few miles down the coast or upriver, a process that further reduced the diversity at each population split.

Because the modern humans of 50,000 years ago were a tropical species, the first people to leave Africa probably crossed the southern end of the Red Sea and kept to roughly the same latitude, hugging the coast until they reached Sahul, the Ice Age continent that then included Australia, New Guinea and Tasmania. The earliest known modern human remains outside Africa, about 46,000 years old, come from Lake Mungo in Australia.

The modern human exodus from Africa occurred at a time when the Pleistocene Ice Age had another 40,000 years to run. To begin with, hunter-gatherer bands were probably stretched out through a strip of mostly tropical climates from northeast Africa to India to Australia. To judge by the behavior of modern hunter-gatherers, these little groups would have been highly territorial and aggressive toward neighbors. To get away from one another and find new territory, bands started moving north into the cold forests and steppes of Europe and East Asia.

The evolutionary pressures for change on these small isolated groups would have been intense. Those migrating eastward faced new environments. Living by hunting and gathering, they would have had to relearn how to survive in each new habitat. The groups moving northward from the equatorial zone of the first migration would have encountered particularly harsh pressures. The last ice age did not end until 10,000 years ago. The first modern humans who moved northward had to adapt to conditions very different from those of their tropical homeland and develop new technologies, such as making tightly fitting clothes and storing food for the winter months. The climate was far colder, the seasonal differences were more pronounced and the problems of keeping warm and finding sustenance during the winter months were severe.

If these obstacles were not daunting enough, the people moving northward encountered armed opposition as well. An earlier wave of humans had left Africa some 500,000 years before and now occupied the Eurasian continent. These humans, called archaic to distinguish them from modern people, included the Neanderthals in Europe and *Homo erectus* in East Asia. Both disappeared about the time that modern humans entered their territories. In the case of the Neanderthals, the archaeological record makes clear that the area of their settlements steadily shrank as that of the modern humans increased, implying that the moderns drove the Neanderthals to extinction. The record from East Asia is not yet detailed enough for the fate of *Homo erectus* to be understood, but a strong possibility is that the species met the same fate as the Neanderthals.

After the occupation of Eurasia, the single gene pool that existed among the small group that left Africa was now fragmented into many different pools. The vast terrain across which humans were now spread, from southern Africa to Europe, Siberia and Australia, prevented any substantial flow of genes between them. Each little population started to accumulate its own set of mutations in addition to those inherited from the common ancestral population. And in each population the forces of natural selection and drift worked independently to process these mutations, making some more common and eliminating others.

If marriage partners had been exchanged freely throughout the human population as it dispersed around the globe, races would never have developed. But the opposite was the case. People as they spread out across the continents at the same time fragmented into small tribal groups. The mixing of genes between these little populations was probably very limited. Even if geography had not been a formidable barrier, the hunter-gatherer groups were territorial and mostly hostile to strangers. Travel was perilous. Warfare was probably incessant, to judge by the behavior of modern hunter-gatherers. Other evidence of warfare lies in the slow growth of the early human population, which was far below the natural birth rate and could imply a regular hemorrhage of deaths in battle.

Once the available territory had been occupied, people overwhelmingly lived and died in the region where they were born. The fact that people were pretty much locked within their home territories until modern times is one of the surprises that has come out of the genome. Several lines of evidence point to this conclusion. All men carry copies of the same original Y chromosome, which, as mentioned above, became universal early in modern human evolution. But mutations started to accumulate in the Y, and each mutation forms a branch point on the human family tree between the men who have it and the men who don't. The root of this branching tree lies in Africa, and its limbs extend around the world in a pattern that follows the path of human migrations. There is not a lot of tangling between the branches, showing that the world filled up in an orderly way and that once it had done so, people then stayed put.

The same story is told by the mitochondrial DNA in tracking the migration of women. More recently, geneticists have been able to survey populations using devices called gene chips that sample the whole genome and provide a much more detailed picture. The gene chips are arrays of short lengths of DNA chosen to recognize half a million sites along the human genome where the sequence often varies. (The variable sites, known as SNPs, or "snips," tell where people differ; the sites on the genome where everyone has the same DNA unit are uninformative.) Because two pieces of DNA will link up chemically if their sequence of bases is exactly complementary* to each other, each short piece of DNA on a gene chip in effect interrogates the genome being tested, saying, "Do you have an A at this site or not?" Thus an entire genome can be scanned to test its sequence at sites that are known to vary from one population to another.

Using a 500,000 snip chip, researchers at Stanford University have found a strong correspondence between the genetics and geographical origins of Europeans. In fact, 90% of people can be located to within 700 kilometers (435 miles) of where they were born, and 50% to within 310 kilometers (193 miles). Europeans are fairly homogeneous at the genetic level, so it is quite surprising that enough genetic differences exist among them to infer a person's origin so precisely.[11]

Another group of researchers looked at Europeans in isolated regions who weren't likely to move much. One site was a Scottish island, another a Croatian village and the third an Italian valley. Anyone who didn't have all four grandparents living in the same

---

*Complementary here means that two strands of DNA carry sequences of DNA units that match each other at each pair of bases. Where one strand has A, the other has T at the same position, and where one strand has G, the other has C. Two such strands have a high chemical affinity for each other, which is weakened if even one pair of bases is not complementary.

region was excluded. Under these conditions, the researchers found they could map individuals to within 8 to 30 kilometers (5 to 19 miles) of their village of origin.

The finding shows that the world's human population is very finely structured in each geographic region in terms of its genetics, with human genomes changing recognizably every few miles across the globe. Such a situation exists only because, until the past few decades, most people have taken marriage partners from very close to where they were born. Such a high degree of local marriage "was probably the norm in rural Europe due to lack of transport or economic opportunities," the researchers conclude.[12]


## Evolutionary Stresses

Once the human population had spread out across the globe, it was subject to a variety of strong evolutionary stresses in the form of a radical makeover of human social organization and population movements that swept over the original settlement pattern. These population shifts were caused by climate change, the spread of agriculture and warfare.

A clue to major population movements after the exodus from Africa is provided by human skin color, which evolved to be dark in equatorial latitudes and pale in northern ones. If one could look at the global population of 25,000 years ago, its differentiation might have been much simpler to trace. Agriculture hadn't yet been invented, and population growth had not yet seriously upset the social structure of small hunter-gatherer groups. Anyone who could have flown around the globe would have seen dark-skinned people inhabiting its equatorial belt, pale-skinned people in its high northern latitudes and a smooth gradation of skin color between them.

What fractured this smooth pattern of association between skin color and latitude? By 25,000 years ago, the Pleistocene Ice Age was nearing its end but was by no means exhausted. The glaciers advanced south one more time, causing the extra cold period known as the Last Glacial Maximum. For the next 5,000 years or so, most of Europe and northern Siberia became uninhabitable. The light-skinned people living in northern latitudes did not wait for the glaciers to bury them. They moved south ahead of the advancing ice fields and as they did so they displaced and probably killed the darker-skinned people to the south of them. The southerners, after all, would hardly have welcomed invasion of their territory and would have defended it to the last. But the northerners would have had the advantage of being genetically and culturally adapted to living in the extreme cold that accompanied them south. Moving ahead of the glaciers, they would have experienced a cooling environment to their liking but arduous for the people whom they were able to displace.

In Europe the retreating northerners found refuges from the cold in Spain and southern France. When the glaciers retreated, starting around 20,000 years ago, both Europe and East Asia were repopulated by former northerners who had survived the Last Glacial Maximum in the southern refuges they had wrested from their previous inhabitants. In this way both Europe and East Asia came to be populated by people with pale skins, the descendants of those who had once lived in the high north.

Another two continents fell into the possession of the pale-skinned northerners around 15,000 years ago, when conditions became warm enough for people living in Siberia to inhabit Beringia, the now sunken landmass that once connected Siberia to Alaska. Perhaps as sea levels rose, some of the inhabitants of Beringia crossed over to Alaska. From there, once the ice sheets melted to open a corridor, they migrated southward to colonize the two continents of North and South America.

Also around 15,000 years ago, there began another process that marked a profound step in the evolution of human social structure—the emergence of the world's first permanent settlements. These appeared independently in Europe, East Asia, Africa and the New World. For the previous 185,000 years, ever since modern humans first appeared in the archaeological record, they had lived as hunters and gatherers. Now, for the first time, people were able to settle down in permanent communities, construct shelters and accumulate property.

The decision to settle cannot have been in any way simple or a matter of pure volition, or it would have taken place many millennia previously. Most likely a shift in social behavior was required, a genetic change that reduced the level of aggressivity common in hunter-gatherer groups. The human fossil record shows that in the period prior to settlement, there had been a gradual thinning of the human skeleton, a process known to physical anthropologists as gracilization. Gracilization typically occurs in the skeletons of wild animal species as they become domesticated. It seems that humans underwent a similar lightening of their bone structure for the same reason—that they were becoming less aggressive. Like animals undergoing domestication, humans shed bone mass because extreme aggressivity no longer carried the same survival advantages, and the most bellicose members of a society were perhaps killed or ostracized. This profound change in social behavior was a necessary precursor to settling down in large communities and learning to get along with people who were not close relations.

The first of these settled societies was the Natufian culture of the Near East, which appears in the archaeological record some 15,000 years ago. Several thousand years after the first settlements, people found themselves inventing agriculture—somewhat inadvertently, because the process of harvesting wild grasses automatically selected for strains more suitable to agriculture. As the climate warmed toward the end of the Pleistocene Ice Age some 10,000 years ago, the incipient systems of agriculture took off, centered on wheat and barley in the Near East and on millet and then rice in China. With the new and more abundant sources of food, population started to increase, and the new farmers expanded their territories. Increased population enhanced social stratification and disparities of wealth within societies, and a brisker tempo of warfare among them. Human social behavior had to adapt to a succession of makeovers as settled tribes developed into chiefdoms, chiefdoms into archaic states and states into empires.

These population expansions vastly changed the pattern of human distribution around the globe. Linguists like to distinguish between what they call mosaic zones and spread zones. The most spectacular mosaic zone still in existence is that of New Guinea. The thickly forested territory is occupied by people who, when discovered by Europeans, were using Stone Age technology and embroiled in endemic warfare. The island's population is separated by territory and by culture. Every 5 to 10 miles, a different language is spoken; the island is home to some 1,200 languages, one-fifth of the world's total. Language is seen as a badge of identity and is deliberately made as different as possible from that of neighboring tribes. Until warfare was supressed by colonial administrators, most New Guineans could not safely travel beyond their native valley.

In contrast to New Guinea with its 1,200 different languages, the United States is a spread zone, because a single language has been spoken from one coast to another since English speakers conquered the original inhabitants with their mosaic zone of many different languages. Much the same kind of process has probably operated throughout the past 50,000 years in a cycle between mosaic zones and spread zones.

When the world outside Africa was first occupied, it would have crystallized, much like the New Guinea linguistic mosaic zone, into many thousands of territories, each occupied by a single tribe. With the passage of time, the language of each tribe would have become more unique and less like that of its neighbors, and its genetics too would have become more distinctive. In each small tribe, different alleles would have drifted up in frequency to fixation or down to extinction.

Why then isn't the global human population far more varied than it is? Because most of these small tribes were destroyed or absorbed into larger tribes as spread zones, propelled by demographic expansion or conquest, rolled like a wave over vast areas of mosaic zone. In Europe, for instance, people bringing the new farming technology from Anatolia, the region now known as Turkey, created a vast spread zone as they overwhelmed the existing hunter-gatherer populations, in part by conquest, in part by intermarriage. An alternative hypothesis is that the spread zone was created by conquest, not the spread of agriculture, as warlike pastoralists from the Russian steppe burst out from their homeland and across Europe and India. In either case, the spread zone reflects the expansion of people who spoke an ancestral tongue, Indo-European, and whose descendants now speak the many languages of the Indo-European family, from Icelandic and Spanish to Iranian and Hindi.

In the Far East too, the rice farmers started to expand, killing the neighboring populations or absorbing them through sheer pressure of numbers. The rise of the Han Chinese to become the world's largest population began just 10,000 years or so ago, this being the time when Mongoloid-type skulls first appear in the archaeological record. The demographic spread of the Han Chinese is still under way, with less numerous neighbors like Tibetans and Uigur Turks finding themselves steadily absorbed into the Han demographic imperium. In Africa, the Bantu expansion is another instance of a spread zone formed by an agriculturally driven population increase. Many of today's races and ethnic groups were probably once small tribes that expanded through population increase, followed by conquering and absorbing outnumbered peoples.

All these evolutionary and historical processes took place independently in each continental population, since there was little flow of people or genes among them. Many salient changes in social behavior—the transition to settled life, the increasing social complexity from village to empire—as well as the engulfment of smaller populations by larger ones, were parallel developments on each continent, although they took place on a different schedule. The first known settlements were in the Near East, followed by those of China, Africa and the Americas. The difference in timing probably depended on population. The denser the population on each continent, the greater the pressure for settlement and the emergence of larger social groups.

Because the genes underlying social behavior are for the most part unknown, the parallel and independent evolution of such genes in the various races cannot yet be demonstrated. But the parallel development of another trait, that of pale skin in East Asians and Europeans, as described below, can now be tracked at the level of the relevant genes.

# A Three Way Split

The emigration from Africa marked the first known major division in the modern human population, between those who remained in Africa and those who left. After the split, the two populations no longer shared a common gene pool, being sharply separated by geography. Migrations back into Africa occurred later, but the numbers of people were far too small to remix the gene pool. Those outside Africa and those within continued to evolve but along different pathways as each adapted to its special set of circumstances.

The next major fork in the human family tree occurred between the populations that colonized the two major halves of the Eurasian continent. Migrants to the north became the ancestors of Caucasians in the west and of East Asians in the east. Caucasians include Europeans, Middle Easterners and the people of the Indian subcontinent. The term Caucasian is avoided by some anthropologists because Blumenbach, who invented the term, believed the inhabitants of the Caucasus were the world's most beautiful people. But Blumenbach, as noted earlier, did not believe Caucasians were superior to other races. Because there is no other word to refer to this important grouping of populations, many geneticists use this term.[13] The date of the split between East Asians and Caucasians is still uncertain but may have been as long as 30,000 years ago.

Both Caucasians and East Asians have light skin, an adaptation to living in high northern latitudes. The default state of primate skin is pale: chimpanzees, under their fur, have white skin (although their faces are dark because of heavy suntan). When our distant ancestors lost their fur, probably because bare skin allowed better sweating and heat control, they developed dark skin to protect a vital chemical known as folic acid from being destroyed by the strong ultraviolet light around the equator. The first modern humans who migrated to the northern latitudes of Europe and Asia were exposed to much less ultraviolet light—too little, in fact, to synthesize enough vitamin D, for which ultraviolet light is required. Natural selection therefore favored the development of pale skin among people living in high northern latitudes. Pale skin may also have been prized in sexual partners, in which case sexual selection, as well as the need to synthesize vitamin D, would have speeded the spread of the necessary alleles. Objectively speaking, pale skin is no more attractive than any other shade. If anything, it is probably less so, to judge by the existence of tanning salons. It could have been prized for arbitrary reasons or, given its association with vitamin D synthesis, because its owners had healthier children in extreme northern latitudes.

Pale skin evolved independently in the Caucasian and East Asian populations, showing that the two populations have remained substantially separate since their split. This is known to be the case because pale skin in Caucasians is caused by a largely different set of genes than those that cause pale skin in East Asians. The independent but parallel evolution of pale skin in the two halves of the Eurasian continent came about because each was exposed to the same stress—the need to protect vitamin D synthesis in northern latitudes. But natural selection can work only with whatever alleles—the different versions of a gene—are present in a population. Evidently different alleles for making pale skin were available in the Caucasian and East Asian populations. This is not so surprising. Making, packaging and distributing the granules of pigment that give skin its color is a complex process, and there are many ways it can be tweaked so as to yield a particular outcome.

Among Africans, dark skin is maintained by the gene known as MC1R. A single version of this gene is found throughout Africa, whereas at least 30 alleles, all different from the African allele, are found among Europeans, and other variants are special to East Asians. It seems that any mutations or changes in the African allele of MC1R lead to lighter skin, which is harmful in the African context. Carriers of such an allele in Africa have no or fewer children, and the variant versions of the MC1R gene that keep cropping up because of mutation are constantly eliminated by purifying selection.[14]

Europeans have pale skin in part because purifying selection on their MC1R gene has been relaxed. But this is not the only reason. They have several alleles that promote pale skin. One is an allele of the gene known as SLC24A5. The SLC24A5 gene specifies a large protein—a chain of amino acid units—in which the 111th unit is the amino acid known as alanine. This is the ancestral form of the gene and is the allele found in almost all Africans and East Asians. Almost all Europeans have an allele in which there is a critical difference in the triplet of three consecutive DNA bases, known as a codon, that specifies the 111th amino acid unit. Different codons determine the 20 kinds of amino acid unit of which proteins are composed. In the case of the SLC24A5 gene, the 111th codon in the ancestral allele is the triplet of bases ACA, which specifies the amino acid alanine. In Europeans, the first A in the triplet has mutated to a G, giving the sequence GCA, and this codon specifies the amino acid known as threonine. This single switch of amino acids alters the function of the protein.

Almost all Europeans have two copies (one from each parent) of the threonine-denoting, skin-lightening allele of the SLC24A5 gene. Africans have two copies of the alanine allele that darkens the skin. African Americans and African Caribbeans who have one copy of each allele have intermediate hues of skin.[15]

East Asians have skin that can be just as pale as that of Europeans. But East Asians carry the ancestral dark skin form of SLC24A5. Natural selection has found other routes by which to lighten the skin of East Asians.

Several other differences are already known between East Asians and Europeans, testifying to the ancient split between the populations. One is the greater thickness of East Asian hair. Africans and Europeans, who have thin hair shafts, carry the same version of a gene called EDAR. A different allele is widespread in East Asians, occurring in 93% of Han Chinese, about 70% of people in Japan and Thailand, and in 60 to 90% of Native Americans. In the 370th codon of the gene, a T has mutated to C, so that the amino acid coded for is alanine instead of valine.[16] Because of the switch of valine (V) to alanine (A) at the 370th codon, the allele is called EDAR-V370A.

East Asians who carry the EDAR-V370A allele also have thick and lustrous hair. But correlation is not proof, so how can one be certain that EDAR-V370A is indeed the cause of East Asians' thick hair shafts? Researchers who wished to prove this point recently generated a strain of mice whose EDAR gene was converted to the East Asian form. They found the mice had thicker hair, proving that the allele is the cause of thick hair in East Asians, but also noticed two other interesting changes.[17]

First, the mice had more eccrine sweat glands than usual in their foot pads. Sweat glands come in two versions, eccrine glands, which secrete water so as to cool the body by evaporation, and apocrine glands, which secrete proteins and hormones. Checking in a Chinese population, the researchers found that the EDAR-V370A causes people too to carry significantly more eccrine glands, a fact that had been previously unknown.

The mice also had smaller breasts than usual, indicating that the EDAR-V370A allele is probably the reason why East Asian women tend to have smaller breasts than African and European women.

A fourth probable effect of EDAR-V370A is that it causes the characteristic dentition of East Asians, whose front teeth look shovel shaped when seen from the back. The mice were less useful in elucidating this effect because their teeth are so different from human teeth.

It may seem surprising that a single gene can have so many profound effects. EDAR has great influence on the body because it is switched on early in embryonic development and helps shape organs such as the skin, teeth, hair and breasts.

The fact that the EDAR-V370A allele has so many effects in East Asians raises the intriguing question of which particular effect was the target of the natural selection that made the allele so common. One possibility is that thick hair and small breasts were much admired by Asian men, or equally that thick hair in either sex was attractive to the other. In either case, these traits would have been acting as agents of sexual selection, a particularly potent form of natural selection.

Another possibility is that the sweat glands were the driving force behind the rise of EDAR-V370A. East Asians are usually assumed to have evolved in a cold climate because of certain traits, such as narrow nostrils and a fold of fat over the eyelid, which seem helpful in conserving body heat. But researchers have calculated that the EDAR variant emerged some 35,000 years ago, at which time central China was hot and humid.

A third possibility is that many or all of the effects of EDAR-V370A were advantageous at one time or another, and that natural selection favored each in turn. Effects of less obvious advantage, such as the shaping of the teeth, were dragged along in the wake of the traits found favorable by natural selection.

EDAR-V370A explains a substantial part, but not all, of the physiological differences between East Asians and other races. Another feature that distinguishes most East Asians from Europeans and Africans has to do with earwax. This substance comes in two forms, wet and dry. The switch between the two types is controlled by two alleles of the gene ABCC11. The allele that causes dry earwax is very common in East Asia. Among the northern Han Chinese and Koreans, 100% of people have the dry allele. The percentage drops to 85% among the southern Han and to 87% in Japan.[18]

Almost all Europeans and all Africans have the wet earwax allele of the ABCC11 gene. This sharp differentiation of the two alleles implies a strong selection pressure. But the function of earwax, like flypaper, is merely to deter insects from crawling into the ear. It seems unlikely that so minor a role would be critical to survival. But as it happens, the two alleles of the ABCC11 are also involved in the apocrine sweat glands.

Unlike the eccrine sweat glands mentioned above, which are found all over the body and secrete just water, the apocrine glands in humans are restricted after birth to just the armpits, nipples, eyelids and other special niches. They make slightly oily secretions, and the specialty of those in the ear is to secrete earwax. The apocrine gland secretions are odorless at first but produce body odor after being decomposed by the bacteria ubiquitous on the skin.

East Asians with the dry earwax allele of the gene produce fewer secretions from their apocrine glands and as a result have less body odor. Among people spending many months in confined spaces to escape the cold, lack of body odor would have been an attractive trait and one perhaps favored by sexual selection.

Yet another East Asian characteristic is the type of skull known to physical anthropologists as Mongoloid. Mongoloid skulls have fine features, a broad head shape, and flattened faces. They also have a distinctive dentition. Africans and Europeans have the same kind of generic human teeth, which is evidently the ancestral pattern. In the East a new tooth pattern emerged, called sundadonty after Sunda, the Ice Age continent that disintegrated after the rise of sea level into Malaysia and the islands of Indonesia. Southeast Asians and the populations derived from them in Polynesia are sundadonts. Some 30,000 years ago, a variation of sundadonty appeared called sinodonty, in which the upper incisors are shovel shaped and some molar teeth have extra roots. Northern Chinese, Japanese and Native Americans, who are descended from Siberian populations, are all sinodonts.

Politically oriented scientists often proclaim that there are no distinct human races, seeking to imply, without actually saying so, that races do not exist. One reason that races exist, though not distinctly, is that the features characteristic of a race are often distributed along a gradient. Almost all northern Chinese have the sinodont pattern of dentition, but the farther one goes toward southern China and Southeast Asia, the greater the percentage of people who are sundadonts and the fewer who are sinodonts. The dry earwax allele is almost universal in northern China but yields to the wet allele toward the south. Most East Asians have the dry earwax gene, but not all do. Most, but not all, have the EDAR-V370A allele.

All these differences are variations superimposed on the common human theme. Even small differences in appearance can be of great social significance, given the strong human tendency to distinguish between the in-group and the out-group. Like the minor variations of language known as dialects, variations in skin or hair color can form the basis on which one group distinguishes itself from its neighbors. If intermarriage then ceases to occur across this fault line, other differences will accumulate, pushing human populations toward differentiation and away from remixing into a common genetic pool.

## The Five Continental Races

Those who assert that human races don't exist like to point to the many, mutually inconsistent classification schemes that have recognized anywhere from 3 to 60 races. But the lack of agreement doesn't mean that races don't exist, only that it is a matter of judgment as to how to define them. As with any species that evolves into geographically based races, there is usually continuity between neighboring races because of gene exchange between them. Because there is no clear dividing line, there are no distinct races—that is the nature of variation within a species. Nonetheless, useful distinctions can be made.

The first step in making sense of human variation and the emergence of races is to follow the historical succession of major population splits. As noted above, the first such split occurred when a small group of people left northeast Africa some 50,000 years ago and populated the rest of the world. The first major division in the human population is thus between Africans and non-Africans. (Africans here denotes people who live south of the Sahara, because those north of the Sahara are largely Caucasian.) Among the non-Africans, there was an early division, whose nature is still poorly understood, between Europeans and East Asians. This gives a three-way split in the human population that corresponds robustly to the three racial groups that everyone can identify at a glance, those of Africans, East Asians and Caucasians. The fact that other peoples may not be so easy to classify does not alter the validity of these three basic categories.

The first migration out of Africa, the one that gave rise to both Europeans and East Asians, eventually reached Sahul, the ancient Ice Age continent that was split by rising sea levels into the three landmasses of Australia, New Guinea and Tasmania. Australian aborigines, surprisingly, turn out to be a race unlike any other. They and their relatives in New Guinea have no trace in their genome of admixture with other races until the historical period. This implies that once Sahul was settled, some 46,000 years ago, the residents fought off all later migrations until the arrival of Europeans in the 18th century. Australian aborigines can reasonably be considered a race, although a minor one in terms of population size, because of their distinctness, antiquity and the fact that they inhabit a continent.

American Indians, the original inhabitants of North and South America, can also be considered a race. Their ancestors were Siberians who originally crossed into Alaska some 15,000 years ago, but American Indians have diverged considerably since then.

A practical way of classifying human variation is therefore to recognize five races based on continent of origin. These are the three principal races—Africans, East Asians and Caucasians—and the two other continent-based groups of Native Americans and Australian aborigines (including the people of New Guinea, an island joined to Australia until the end of the last ice age).

At the land boundaries where races meet, there are often intermarried or admixed populations, as geneticists call them. Palestinians, Somalis and Ethiopians, for instance, are admixtures of African and Caucasian populations. The Uigur Turks of northwestern China and the Hazara of Afghanistan are admixtures of Caucasian and East Asian populations. African Americans are an admixture mostly of Africans and Caucasians.

Within each continental race are smaller groupings which, to avoid terms like subrace or subpopulation, that might be assumed to imply inferiority, may be called ethnicities. Thus Finns, Icelanders, Jews and other groups with recognizable genetics are ethnicities within the Caucasian race.

Such an arrangement, of portioning human variation into five continental races, is to some extent arbitrary. But it makes practical sense. The three major races are easy to recognize. The five-way division matches the known events of human population history. And most significant of all, the division by continent is supported by genetics.

# 5

———•◆•———

# THE GENETICS
# OF RACE

Selfish and contentious people will not cohere, and without coherence nothing can be effected. A tribe rich in the above qualities would spread and be victorious over other tribes: but in the course of time it would, judging from all past history, be in its turn overcome by some other tribe more highly endowed. Thus the social and moral qualities would tend slowly to advance and be diffused throughout the world.

—CHARLES DARWIN[1]

In the case of human races, the genetic differences from one race to another are slight and subtle. One might expect that different races would have different genes, but they don't. All humans, so far as is known, have the same set of genes. Each gene comes in various alternative forms, called alleles, so the next expectation might be that races would be distinguished by having different alleles of various genes. But this too is not how the system works. There are a mere handful of known cases where a particular allele of a gene occurs in only one race.

The genetic differences between human races turn out to be based largely in allele frequencies, meaning the percentages of each allele that occur in a given race. How a mere difference in allele frequencies could lead to differences in physical traits is explained below.

## Races as Clusters of Variation

A necessary approach to studying racial variation is to look not for absolute differences but at how the genomes of individuals throughout the world cluster together in terms of their genetic similarity. The result is that everyone ends up in the cluster with which they share the most variation in common. These clusters always correspond to the five continental races in the first instance, though when extra DNA markers are used, the people of the Indian subcontinent sometimes split away from Caucasians as a sixth major group, and people of the Middle East as a seventh.

One of the first genetic clustering techniques depended on examining an element of the genome called tandem repeats. There are many sites on the genome where the same pair of DNA units is repeated several times in tandem. CA stands for the DNA unit known as a cytosine followed by adenine, so the DNA sequence CACACACA would be called a tandem CA repeat. The string of repeats occasionally confuses the DNA copying apparatus, which every few generations may add or drop a repeat unit during the copying process that has to occur before a cell can divide. Sites at which repeats occur therefore tend to be quite variable, and this variability is useful for comparing populations.

In 1994, in one of the earliest attempts to study human differentiation in terms of DNA differences, a research team led by Anne
Bowcock of the University of Texas and Luca Cavalli-Sforza of Stanford University looked at CA repeats at 30 sites on the genome in
people from 14 populations. Comparing their subjects on the basis
of the number of CA repeats at each genomic site, the researchers
found that people clustered together in groups that were coincident
with their continent of origin. In other words, all the Africans had
patterns of CA repeats that resembled one another, all the American
Indians had a different pattern of repeats and so on. Altogether there
were 5 principal clusters of CA repeats, formed by people living in
each of the 5 continental regions of Africa, Europe, East Asia, the
Americas and Australasia.[2]

Many larger and more sophisticated surveys have been done
since, and all have come to the same conclusion, that "genetic differentiation is greatest when defined on a continental basis," writes Neil
Risch, a statistical geneticist at the University of California, San
Francisco. "Effectively, these population genetic studies have recapit-
ulated the classical definition of races based on continental ancestry—
namely African, Caucasian (Europe and Middle East), Asian, Pacific
Islander (for example, Australian, New Guinean and Melanesian),
and Native American."[3]

In one of these more sophisticated studies, a team led by Noah
Rosenberg of the University of Southern California and Marcus Feldman of Stanford University looked at the number of repeats at 377
sites on the genome of more than 1,000 people around the world.
When this many sites are examined on a genome, it's possible to
assign segments of an individual's genome to different races if he or
she has mixed ancestry. This is because each race or ethnicity has a
characteristic number of repeats at each genomic site.

The Rosenberg-Feldman study showed, as expected, that the

1,000 individuals in their study clustered naturally into five groups, corresponding to the five continental races. It also brought out the fact that several Central Asian ethnicities, such as Pathans, Hazara and Uigurs, are of mixed European and East Asian ancestry. This is not a surprise, given the frequent movement of peoples to and fro across Central Asia.

Language is often an isolating mechanism that deters intermarriage with neighboring groups. The Burusho, a people of Pakistan who speak a unique language, turn out also to be unlike their neighbors genetically. Within races, the Rosenberg-Feldman study showed that different ethnicities could be recognized. Among Africans, it is easy to distinguish by their genomes the Yoruba of Nigeria, the San (a click-speaking people of southern Africa) and the Mbuti and Biaka pygmies.

Many populations are not highly mixed, and the Rosenberg-Feldman survey confirmed the remarkable extent to which people throughout history have lived and died in the place where they were born.[4]

In the ancestral human population in Africa, a large number of alleles had developed for each gene over many generations. Those who migrated out of Africa took away only a sample of these alleles. And each time a new group split off, the number of alleles from the original population again decreased.

The farther away from Africa that this process continued, the less was the diversity of alleles. This downhill gradient happens with any population that expands too far from its origins to maintain the regular interbreeding that keeps the gene pool well mixed.

A genetic gradient, or cline, is what some researchers prefer to think exists in place of races. "There are no races, there are only clines," asserted the biological anthropologist Frank Livingstone.[5] Critics raised the same objection against the Rosenberg-Feldman result, alleging that the clustering of individuals into races was an artifact and that with a geographically more uniform sampling approach, the researchers would have seen only clines.[6] The Rosenberg-Feldman team then reanalyzed their data and gave their survey finer resolution by looking at 993 sites, not just 377, on each of the genomes in their study. They found that the clusters are real. Although there are gradients of genetic diversity, there is also a clustering into the continental groups described in their first article.[7]

Rosenberg and Feldman compared people's genomes on the basis of DNA repeats. Another kind of DNA marker has since become available for global population comparison—the SNP, which is more useful for medical studies. SNP stands for single nucleotide polymorphism, meaning a site on the genome where some people have a different kind of DNA unit from that of the majority. A vast preponderance of sites on the genome are fixed, meaning everyone has the same DNA unit, whether A, T, G or C. The fixed sites, being all the same, say nothing about human variation. It's the SNP sites, which are variable, that are of particular interest to geneticists because they afford a direct way of comparing populations. To exclude the many random mutations that occur just in particular individuals and have no wider importance, SNPs are arbitrarily defined as sites on the genome where at least 1% of the population has a DNA unit other than the standard one.

A research group led by Jun Z. Li and Richard M. Myers has applied a clustering program like that used by Rosenberg and Feldman to almost 1,000 people in 51 populations across the globe. Each person's genome was examined at 650,000 SNP sites. On the basis of SNPs, just as with the DNA repeats, people sampled from around the world clustered into 5 continental groups. But in addition, the SNP library brought to light two other major clusters. These had not emerged in the Rosenberg-Feldman study, which had used fewer markers. The more DNA markers that are used, whether tandem repeats or SNPs, the more subdivisions can be established in the human population.

One of the new clusters is formed by the people of Central and South Asia, including India and Pakistan. The second is the Middle East, where there is considerable admixture with people from Europe and Africa.[8] It might be reasonable to elevate the Indian and Middle Eastern groups to the level of major races, making seven in all. But then many more subpopulations could be declared races, so to keep things simple, the five-race, continent-based scheme seems the most practical for most purposes.

Within each continental race, the SNP analysis could separate out further subgroups. Within Europe it distinguished French, Italians, Russians, Sardinians and Orcadians (people who live in the Orkney Islands, north of Scotland). In China the northern Han can be distinguished from the southern Han.

Groupings within Africa are of particular interest because this is where modern humans spent the first 150,000 years of their existence. In the most thorough survey of Africa so far, Sarah Tishkoff and colleagues surveyed people from 121 populations, scanning their genomes at 1,327 variable sites, most of them DNA repeats. The survey brought to light 14 different ancestral groups within Africa. Tishkoff found that, unlike in the rest of the world, where there are definable continental races, in Africa most populations are admixtures of several ancestral groups. There have presumably been a larger number of migration events within Africa, which served to mix up populations that were originally separate. The most recent large-scale migration was the Bantu expansion, a population explosion driven by new agricultural technology. Within the past few thousand years, Bantu speakers from the region of Nigeria and Cameroon in West Africa have migrated across to eastern Africa and down both coasts to southern Africa. Only a few groups have kept relatively clear of the churning of populations within Africa. These include the click-speaking peoples of Tanzania and southern Africa, who until recently have been hunter-gatherers, and the various pygmy groups, who live deep in the forest.[9]

The click-speakers and pygmies may be remnants of a much earlier hunter-gatherer population that once occupied a large part of southern Africa and the eastern coast as far north as Somalia. The click-speakers speak a group of languages known as Khoisan, which are unlike any others and have only very distant relationships among themselves, probably reflecting their great antiquity. The pygmy groups too may once have spoken Khoisan languages but it is impossible to know for sure, because they have lost their original languages.

Africa has four language superfamilies, of which Khoisan is one and the other three are Niger-Kordofanian (also known as Niger-Congo), Nilo-Saharan and Afro-Asiatic. The Niger-Kordofanian languages, the most widespread, were carried from western to eastern Africa and then south by the Bantu expansion, a great stream of migrations from the proto-Bantu homeland in western Africa that began in about 1000 BC and reached southern Africa a thousand years later. Afro-Asiatic languages are spoken in a broad belt across northern Africa, and the Nilo-Saharan speakers are sandwiched between Afro-Asiatic to the north and Niger-Kordofanian to the south.

Genetics generally correlates with language family, except in the case of populations that have switched languages; the pygmies now speak Niger-Kordofanian languages, and the Luo of Kenya, whose genetics place them with Niger-Kordofanian speakers, now speak a Nilo-Saharan language.

The Tishkoff team surveyed African Americans from Chicago, Baltimore, Pittsburgh and North Carolina and found that 71% of their genomes, on average, matched the genetics of Niger-Kordofanian speakers, 8% matched that of other African populations and 13% were European. These percentages varied greatly from one individual to another.

The origin of a species can often be located by surveying the genetic diversity in its members and seeing where diversity is highest. This is because the founding population will have had longest to accumulate the mutations that generate diversity, and the groups that migrate away will carry with them only a sample of the original mutations. (Other forces, like natural selection, reduce diversity by eliminating harmful mutations and sweeping away others when a beneficial mutation is favored.) On the basis of the new African and other genomic data, the origin of the modern human migration lies in southwestern Africa, near the border of Namibia and Angola, in a region that is the current homeland of the San click-speakers. The finding is not definitive, because the distribution of ancient populations may have been rather different from those of today. Nonetheless, the fact that human genetics points to a single origin confirms that today's races are all mere variations on the same theme.

## Fingerprints of Selection in the Human Genome

Both repeated DNA units and SNPs, the two kinds of DNA marker used by the surveys described above, lie for the most part outside genes and have little or no effect on a person's physical makeup. They are what geneticists call neutral variations, meaning that they are ignored by natural selection. What then is it that makes human populations differ from one another?

Natural selection is the major shaper of differences, especially in large societies. In small societies, genetic drift—the luck of the draw as to which alleles make it into the next generation—can be a significant influence. But natural selection, often in concert with drift, is a major force over the long run. With the advent of fast methods of genome sequencing, geneticists have at last begun to delineate the fingerprints of natural selection in remodeling the human genome. These fingerprints are both recent and regional, meaning that they differ from one race to another.

The regional nature of selection was first made evident in a genomewide scan undertaken by Jonathan Pritchard, a population geneticist at the University of Chicago, in 2006. He looked for genes under selection in the three major races—Africans, East Asians and Europeans (or more exactly Caucasians, but European genetics are at present much better understood, so European populations are the usual subjects of study). Copious genetic data had been collected on each race as part of the HapMap, a project undertaken by the National Institutes of Health to explore the genetic roots of common disease. In each race Pritchard found about 200 genetic regions that showed a characteristic signature of having been under selection (206 in Africans, 185 in East Asians and 188 in Europeans). But in each race, a largely different set of genes was under selection, with only quite minor overlaps.[10]

The evidence of natural selection at work on a gene is that the percentage of the population that carries the favored allele of the gene has increased. But though alleles under selection become more common, they rarely displace all the other alleles of the gene in question by attaining a frequency of 100%. Were this to happen often in a population, races could be distinguished on the basis of which alleles they carried, which is generally not the case. In practice, the intensity of selection often relaxes as an allele rises in frequency, because the needed trait is well on the way to being attained.

Geneticists have several tests for whether a gene has been a recent target for natural selection. Many such tests, including the one devised by Pritchard, rest on the fact that as the favored allele of a

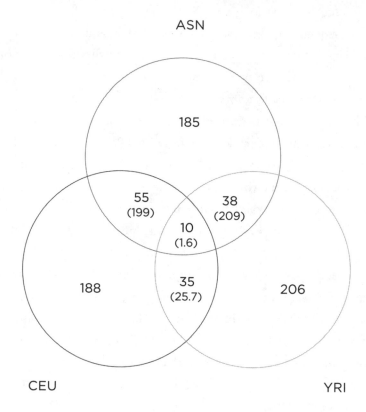

Figure 4.1. Regions of the genome that are highly selected in the three major races. ASN = East Asian, a sample of Chinese and Japanese. YRI = Yoruba, a West African people. CEU = European.

From Jonathan Pritchard, *PLoS Biology* 4(2006):446–58.

gene sweeps through a population, the amount of genetic diversity in and around the gene is reduced in the population as a whole. This is so because increasing numbers of people now carry the same se-quence of DNA units at that site, those of the favored allele. So the result of such a sweep is that DNA differences between members of a population are reduced in the region of the genome affected by the sweep. The concept of using sweeps as signatures of natural selection is discussed further below.

Other researchers too have found that in doing genome scans for the fingerprints of natural selection, each major race or continental population has its own distinctive set of sites where selection has occurred.

These sites of selection are often very large and contain many genes, making it hard or impossible to decide which specific gene was the target of natural selection. In a new approach, which takes advantage of the many whole genomes that have now been decoded, Pardis Sabeti of Harvard and colleagues have defined 412 regions under selection in Africans, Europeans and East Asians. The regions are so small that most contain one or no genes. Those without genes presumably contain a control element, meaning a stretch of DNA that regulates some nearby gene.[11]

Of the 412 regions of the human genome shown to be under selection, 140 were under selection just in Europeans, 140 in East Asians and 132 in Africans.[12] The absence of any overlap, meaning genes selected in two or more populations, as was found by Pritchard, is due to the Sabeti team's genome scanning method, which depended in part on looking for sites at which the three races differed.

Each gene under selection will eventually tell a fascinating story about some historical stress to which the population was exposed and then adapted. A case in point is the analysis of the EDAR-V370A allele which, as described in the previous chapter, is the cause of thick hair and other traits in East Asians. But those narratives are for the moment inaccessible. The exploration of the human genome is so much at its beginning that the precise function of most genes is unknown.

Still, even though the exact tasks of most genes are still uncertain, the general roles of most genes can be inferred by comparing the DNA sequence of any unknown gene with those of known genes recorded in genomic data banks. The known genes are grouped into general functional categories, like brain genes or genes involved in metabolism, and since function is related to structure, the genes in each category have a characteristic sequence of DNA units. By comparing the DNA sequence of any new gene with the data bank sequences, the gene can be assigned to a general functional category. The genes Pritchard identified as shaped by natural selection included genes for fertilization and reproduction, genes for skin color, genes for skeletal development and genes for brain function. In the brain function category, four genes were under selection in Africans and two each in East Asians and Europeans. What these genes do within the brain is largely unknown. But the findings establish the obvious truth that brain genes do not lie in some special category exempt from natural selection. They are as much under evolutionary pressure as any other category of gene.

Population geneticists have developed several different kinds of tests to see if natural selection has influenced the DNA sequence of a gene. All these tests are statistical, and many depend on the disturbance in gene frequencies that is caused as a favored gene sweeps through a population. Natural selection cannot pick out single genes or even single mutations in DNA. Rather, it depends on the process called recombination, in which the mother's and father's genomes are shuffled prior to creating eggs and sperm.

In the egg-making or sperm-making cells, the two sets of chromosomes that a person has inherited, one from their mother and one from their father, are lined up side by side, and the cell then forces them to exchange large sections of DNA. These new composite chromosomes, consisting of some sections from the father's genome and some from the mother's, are what is passed on to the next generation.

The swapped sections, or blocks, may be 500,000 DNA units in length, long enough to carry several genes. So a gene with a beneficial mutation will be inherited along with the whole block of DNA in which it is embedded. It's because beneficial genes lie in such a large block that the effect of natural selection on the genome can be detected—the favored blocks sweep out large regions of the genome as they spread through a population.

Generation by generation, the block of DNA with the favored version of a gene gets to be carried by more and more people. Eventually, the new allele may sweep through the entire population, in which case geneticists say it has gone to fixation. But most sweeps do not carry an allele to fixation because, as already noted, the intensity of selection on a beneficial allele relaxes as the trait is molded toward its most efficient form.

Whether a sweep is complete or partial, the favored blocks of DNA eventually get whittled down over the generations, because the cuts that generate them are not always made in the same places on the chromosome. After just 30,000 years or so, according to one calculation, the blocks get too short to be detectable. This means that most genomewide scans for selection are looking at events that occurred just a few thousand years ago, very recently in human evolution.

Biologists have long had to depend on the evidence from fossils to judge the speed of evolution. But fossils capture just the bones of an animal. And since the skeletal structure of a species changes only slowly, evolution has long seemed a glacially slow and plodding process.

With the ability to decode DNA sequences, biologists can examine the raw programming of evolutionary change and track every gene in a species' repertoire. It's now clear that evolution is no sluggard. There are already clear examples of human evolutionary change within the past few thousand years, such as Tibetans' adaptation to high altitude, starting from just 3,000 years ago. Of course, every gene in the human genome has been intensely shaped by natural selection at one time or another. But with most genes, the selection was accomplished eons before humans or even primates had evolved. The fingerprints of these ancient selection events have long since faded from sight. The type of selection picked up by most genome scans is very recent selection, meaning within the past 5,000 to 30,000 years or so, but fortunately this is a period of great interest for understanding human evolution.

More than 20 scans for selection have now been performed on the human genome. They do not all mark the same regions as being under selection but that is not surprising since the authors use different kinds of tests and different statistical methods, which are in any case imprecise. But if one takes just the regions marked by any two of the scans, then 722 regions, containing some 2,465 genes, have been under recent pressure of natural selection, according to an estimate by Joshua M. Akey of the University of Washington. This amounts to 14% of the genome.[13]

That so much of the genome has been under natural selection strong enough to be detectable shows how intense human evolution must have been in the past few thousand years. A principal driver of evolutionary change would have been the need to adapt to a wide range of new environments. In proof of that point, some 80% of the 722 regions under selection are instances of local adaptation, meaning that they occur in one of the three main races but not in the other two.

The genes under selection affect a large number of biological traits, prominent among them being skin color, diet, bone and hair structure, resistance to disease and brain function.

A similar finding emerged from a particularly comprehensive genome scan conducted by Mark Stoneking and colleagues. Stoneking, a population geneticist at the Max Planck Institute for Evolutionary Anthropology in Leipzig, is known for having developed an ingenious way of estimating when humans first started to wear clothes. The body louse, which lives only in clothes, evolved from the head louse, which lives on hair. Stoneking realized that a date for the first tight-fitting clothes could be derived by using genetic methods to date the birth of the body louse lineage—about 72,000 years ago.[14]

In his genome survey, Stoneking found many genes under selection that affected people's interaction with their environment, such as genes involved in metabolizing certain classes of food and genes that mediate resistance to pathogens. Among the genes under selection he also found several that were involved in aspects of the nervous system, such as cognition and sensory perception.

The genes of the nervous system have been under selection for the same reason as the other genes—to help people adapt to local circumstances. Changes in social behavior may well have been foremost, given that it is largely through their society that people interact with their environment. Signals of selection in brain genes "may be related to how different human groups interact behaviorally with their environment and/or with other human groups," Stoneking and colleagues wrote.[15]

Another regional trend indicated by the genome scans is that there seem to be more genes under selection in the genomes of East Asians and Europeans than in those of Africans. Not all genome scans have reported such a finding—the Pritchard scan described above did not—and African populations have been poorly sampled so far. But in a subsequent scan, Pritchard and others did find evidence for more sweeps outside Africa.

"A plausible explanation is that humans experienced many novel selective pressures as they spread out of Africa into new habitats and cooler climates," they wrote. "Hence there may simply have been more sustained selective pressures on non-Africans for novel phenotypes."[16] Phenotype refers to the physical trait or organism produced by the DNA, as contrasted with the DNA itself, which is known as the genotype. One obvious example of a novel phenotype needed outside Africa is that of skin color. Africans have retained the default dark skin of the ancestral human population, whereas East Asians and Europeans, descendants of populations who adapted to extreme northern latitudes, have evolved pale skin.

Both within Africa and in the world outside, social structure underwent a radical transition as populations began to grow after the beginning of agriculture some 10,000 years ago. Independently on all three continents, people's social behaviors started to adapt to the requirements of living in settled societies that were larger and more complex than those of the hunter-gatherer band. The signature of such social changes may be written in the genome, perhaps in some of the brain genes already known to be under selection. The MAO-A gene, which influences aggression and antisocial behavior, is one behavioral gene that, as mentioned in the previous chapter, is known to vary between races and ethnic groups, and many more will doubtless come to light.

## Hard Sweeps and Soft Sweeps

Textbooks about evolution discuss favorable alleles that sweep through a population and become universal. There are many ancient alleles that have probably become fixed in this way. All humans, at least compared with chimpanzees, carry the same form of the FOXP2 gene, which is a critical contributor to the faculty of speech. A variation called the Duffy null allele has become almost universal among Africans because it was an excellent defense against an ancient form of malaria. A gene called DARC (an acronym for Duffy antigen receptor for chemokines) produces a protein that sits on the surface of red blood cells. Its role is to convey messages from local hormones (chemokines) to the interior of the cell. A species of malarial parasite known as *Plasmodium vivax,* once endemic in parts of Africa, learned how to use the DARC protein to gain entry into red blood cells. A mutated version of the DARC gene, the Duffy null allele, then became widespread because it denies the parasite access to the blood cells in which it feeds and thus provides a highly effective defense. Almost everyone in Africa carries the Duffy null allele of DARC, and almost no one outside does.[17]

Many other mutations have arisen to protect people against current strains of malaria, such as those that cause sickle-cell anemia and the thalassemias. Sickle-cell anemia occurs with high frequency in Africa, and beta-thalassemia is common in the Mediterranean, but neither has attained the universality of the Duffy null allele within a population. Another widespread but fairly exclusive allele is associated with skin color. This is an allele of KITLG (an acronym for KIT ligand gene) which leads to lighter skin. Some 86% of Europeans and East Asians carry the skin lightening allele of KITLG. This allele evolved because of a mutation in the ancestral, skin-darkening version of KITLG, which is carried by almost all Africans.[18] A skin-lightening allele of another gene, called SLC24A5, has swept almost completely through Europeans.

But the number of such genes, in which one allele has gone to fixation in one race and a different allele in another, is extremely small and in no way sufficient to account for differences between populations. Pritchard found no cases of an allele going to fixation among the Yoruba, a large African tribe in Nigeria. This has led him and other geneticists to conclude that complete sweeps have been much rarer in human evolution than supposed.[19]

But given that all humans have the same set of genes and that there have been almost no full sweeps that push different alleles to dominance in different races, how have races come to differ from one another? The answer that has dawned on geneticists in the past few years is that you don't always need a full sweep to change a trait. Many traits, like skin color or height or intelligence, are controlled by a large number of different genes, each of which has alleles that individually make small contributions to the trait. So if just some of these alleles become a little more common in a population, the trait will be significantly affected. This process is called a soft sweep, to contrast it with a full or hard sweep, in which one allele of a gene displaces all the others in a population.

Pritchard gives the example of height, which is affected by hundreds of genes, because there are so many ways in which height can be increased. Suppose there are 500 such genes and each comes in two forms, with one allele having no effect on height and the other increasing it by 2 millimeters. An individual's height depends on how many of the height-enhancing alleles he inherits. And that number in turn is determined by the frequency of each type of allele, meaning how common it is in the population. So if each of the height-promoting alleles becomes just 10% more common in the population, almost everyone will inherit more of them, and the average person's height will increase by 200 millimeters, or 20 centimeters (8 inches).[20]

This soft sweep process—a small increase in frequency in many genes—is a much easier way for natural selection to operate than through the hard sweeps—the major jump in frequency of a single allele—that are often assumed to be the main drivers of evolution. The reason is that the hard sweeps depend on a mutation creating a novel allele of great advantage, which happens only very rarely in a population. In a small population, it may take many generations for such a mutation to occur. Soft sweeps, on the other hand, act on alleles that already exist and simply make some of them more common. Soft sweeps can thus begin whenever they are needed.

So suppose a group of pygmies were to leave their forest habitat and start herding cattle in a hot climate, where it's advantageous to be tall and thin, like the Nuer and Dinka of the Sudan. The pygmies who were slightly taller would produce more children, and the height-promoting alleles of the genes that affect height would imme-diately begin to become more common in the population. In each generation, an individual would have a slightly greater chance of inheriting the height-promoting alleles, and the population would quite quickly become considerably taller.

Consider, on the other hand, a trait in which there is no existing variation, such as the ability to digest milk in adulthood. For most of human existence and still in most people today, the gene for lactase is switched off shortly after weaning. To keep the gene switched on requires a beneficial mutation in the region of promoter DNA that controls it. But the promoter region is some 6,000 DNA units in length and occupies a minuscule fraction of the 3 billion units of the genome. In a small population, it might take many generations for the right mutation to occur in so small a target.

Thus it seems to have taken around 2,000 years—some 80 generations—after the start of cattle breeding for the right mutation in the lactase promoter region to appear among the people of the Funnel Beaker Culture, cattle herders who occupied northern Europe some 6,000 years ago. Once established, the mutation spread rapidly and is now found at high frequency in northern Europe.

Three mutations, which differ from one another and from the European mutation but have the same effect, arose independently among pastoralist peoples in eastern Africa and have swept through roughly 50% of the population. In each case, evidently, evolution has had to wait until the right mutation occurred, whereupon the allele grew more common because of the great advantage it conferred.

In sum, hard sweeps cannot start until the right mutation occurs, and then they may take many generations to sweep through a popu-lation. Soft sweeps, based on standing variation in the many genes that control a single trait, can start immediately. For a species that undergoes a sudden expansion in its range and needs to adapt quickly to a succession of different challenges, the soft sweep is likely to be the dominant mechanism of evolutionary change. This explains why so few hard sweeps are visible in the human genome. Soft sweeps are presumably far more common, though at present are very hard to detect. The reason is the difficulty of distinguishing between the minor changes in allele frequency caused by genetic drift and the also minor changes brought about by natural selection through a soft sweep.

## The Genetic Structure of Race

It is now possible to understand the structure of human variation, at least in broad outline. Different populations don't have different genes—everyone has the same set. Of the traits specific to one race or another, a few are encoded in hard sweep alleles that have gone almost to fixation, such as the Duffy null allele or some of the alleles involved in shaping skin color, but many more are probably encoded in soft sweeps and hence in mere differences in the frequency of the cluster of alleles that shape each trait.

The fact that genes work in combination explains how there can be so much variation in the human population and yet so few fixed differences between populations.

Given the importance of allele frequencies in shaping specific traits, it's not surprising that they afford a means of identifying an individual's race. Excluding subjects of a different race is an essential procedure in surveys to detect the alleles that contribute to complex diseases like diabetes and cancer. The idea of these surveys, known as genomewide association studies, is to see if people who are particularly prone to disease are also more likely to carry a particular allele. If so, the allele may be associated with the disease. But the statistics can be confounded if the population being surveyed includes people of more than one race. An apparent association may emerge between the disease state and a particular allele even though the association is really due to some patients belonging to another race, one that naturally has a high frequency of the allele in question.

Medical geneticists have therefore developed sets of test alleles that can be used to distinguish one race from another. Some alleles, particularly those with large differences in frequency between races, are more useful than others. These race-distinguishing DNA sites are known blandly as AIMs, or ancestry informative markers. Using a set of 326 AIMs, researchers achieved a nearly perfect correspondence between the race that subjects said they belonged to and the race to which they were assigned genetically.[21] A set of 128 AIMs suffices to assign people to their continental race of origin, whether European, East Asian, American Indian or African.[22] (The fifth continental race, Australian aborigines, could doubtless be identified just as easily, but political restrictions have so far largely blocked the study of aborigine genetics.)

With greater numbers of markers, more closely related groups can be distinguished, such as the various ethnicities within Europe.

Some biologists insist that AIMs do not prove the existence of race and that they point instead to geographic origin. But geographic origin correlates very well with race, at least on the continental level.

Apart from genetic markers like the Duffy null allele, found almost exclusively in people of African ancestry, most AIMs are alleles that are just somewhat more common in one race than in another. A single AIM that occurs in 45% of East Asians and 65% of Europeans says that the carrier is a little more likely to be European, but is hardly definitive. When the results from a string of AIMs are combined, however, an answer with high statistical probability is obtained. This is the same general method used in DNA fingerprint-ing, except that the 14 sites at which the genome is sampled in forensic DNA analysis are not SNPs but variable runs of DNA repeats.

The approach of comparing allele frequencies can even be used with people of mixed race to assign component parts of an individual's genome to their parent's racial origin. When people of different races marry, their children are perfect blends of their parents' genes. But at the genetic level, the chunks of DNA that came from the mother's and the father's races remain separate and distinguishable for many generations. Researchers can track along the chromosomes of African Americans, assigning each stretch of DNA to either African or European ancestors. In one recent study, researchers analyzed the genomes of almost 2,000 African Americans and found that 22% of their DNA came from European ancestors and the rest from Africans, a conclusion in line with several previous reports.[23]

The same study found evidence that African Americans may already have begun adapting genetically to the American environment in the several generations since their ancestors arrived in the United States. The malaria-protecting genetic variants common in Africans, such as the variation that causes sickle-cell anemia, are no longer a necessity of survival in the United States, so the pressure of natural selection to retain these variants would be relaxed. The authors found some evidence that these variants have indeed declined in frequency in African Americans, while genes that provide protection against influenza have grown more common. The finding, if confirmed, would be a striking instance of evolutionary change within the past few hundred years.

Over the last 50,000 years, modern humans have been subjected to enormous evolutionary pressures, in part from the consequences of their own social culture. They explored new ranges and climates and developed new social structures. Fast adaptation, particularly to new social structures, was required as each population strove to exploit its own ecological niche and to avoid conquest by its neighbors. The genetic mechanism that made possible this rapid evolutionary change was the soft sweep, the reshaping of existing traits by quick minor adjustments in the sets of alleles that controlled them.

But what began as a single experiment with the ancestral human population became a set of parallel experiments once the ancestral population had spread throughout the world. These independent evolutionary paths led inevitably to the different human populations or races that inhabit each continent.

## Arguments Against the Existence of Race

Readers who are by now persuaded that recent human evolution has resulted in the existence of races may wish to proceed to the next chapter. But for those who remain perplexed that so many social scientists and others should argue race does not exist, here is an analysis of some of their contentions.

Start with Jared Diamond, the geographer and author of *Guns, Germs, and Steel*, who was quoted in chapter 4 as comparing the idea of race with the belief that the Earth is flat. His principal argument for the nonexistence of race is that there are many different "equally valid procedures" for defining human races, but since all are incompatible, all are equally absurd. One such procedure, Diamond proposes, would be to put Italians, Greeks and Nigerians in one race, and Swedes and Xhosas (a southern African tribe) in another.

His rationale is that members of the first group carry genes that confer resistance to malaria and those of the second do not. This is just as good a criterion as skin color, the usual way of classifying races, Diamond says, but since the two methods lead to contradictory results, all racial classification of humans is impossible.

The first flaw in the argument is the implied premise that people are conventionally assigned to races by the single criterion of skin color. In fact, skin color varies widely within continents. In Europe it runs from light-skinned Swedes to the olive complexion of southern Italians. Skin color is thus an ambiguous marker of race. People belong to a race not by virtue of any single trait but by a cluster of criteria that includes the color of skin and hair, and the shape of eyes, nose and skull. It is not necessary for all these criteria to be present: some East Asians, as noted above, lack the EDAR allele for thick hair, but they are still East Asians.

The single criterion that Diamond proposes as an alternative, genes that confer resistance to malaria, makes no evolutionary sense. Malaria became a significant human disease only very recently, some 6,000 years ago, and each race then independently developed resistance to it. Italians and Greeks resist malaria because of mutations that also cause the blood disease known as thalassemia, whereas Africans resist malaria through a different mutation that causes sickle-cell anemia. The trait of resisting malaria is one that has been acquired secondarily to race, so obviously it is not an appropriate way of classifying the populations. A scholar's duty is to clarify, but Diamond's argument seems designed to distract and confuse.

A more serious and influential argument, also designed to banish race from the political and scientific vocabulary, is one first advanced by the population geneticist Richard Lewontin in 1972. Lewontin measured a property of 17 proteins from people of various different races and calculated a measure of variation known as Wright's fixation index. The index is designed to measure how much of the variation in a population resides in the population as a whole and how much is due to differences between specific subpopulations.

Lewontin's answer came out to 6.3%, meaning that of all the variations in the 17 kinds of protein he had looked at, only 6.3% lay between races, while a further 8.3% lay between ethnic groups within races. These two sources of variation add up to around 15%, leaving the rest as common to the population as a whole. "Of all human variation, 85% is between individual people within a nation or tribe," Lewontin stated. He concluded on this basis that "human races and individuals are remarkably similar to each other, with the largest part by far of human variation being accounted for by the differences between individuals."

He went on to say that "Human racial classification is of no social value and is positively destructive of social and human relations. Since such racial classification is now seen to be of virtually no genetic or taxonomic significance either, no justification can be offered for its continuance."[24]

Lewontin's thesis immediately became the central genetic plank of those who believe that denying the existence of race is an effective way to combat racism. It is prominently cited in *Man's Most Dangerous Myth: The Fallacy of Race,* an influential book written by the anthropologist Ashley Montagu with the aim of eliminating race from the political and scientific vocabulary. Lewontin's statement is quoted at the beginning of the American Anthropological Association's statement on race and is a founding principle of the assertion by sociologists that race is a social construct, not a biological one.

But despite all the weight that continues to be placed on it, Lewontin's statement is incorrect. It's not the basic finding that is wrong. Many other studies have confirmed that roughly 85% of human variation is among individuals and 15% between populations. This is just what would be expected, given that each race has inherited its genetic patrimony from the same ancestral population that existed in the comparatively recent past.

What is in error is Lewontin's assertion that the amount of

variation between populations is so small as to be negligible. In fact it's quite significant. Sewall Wright, an eminent population geneticist, said that a fixation index of 5% to 15% indicates "moderate genetic differentiation" and that even with an index of 5% or less, "differentiation is by no means negligible."[25] If differences of 10 to 15% were seen in any other than the human species they would be called subspecies, in Wright's view.[26]

Why should Wright's judgment that a fixation index of 15% between races is significant be preferred over Lewontin's assertion that it is negligible? Three reasons: (1) Wright was one of the three founders of population genetics, the relevant discipline; (2) Wright invented the fixation index, which is named after him; (3) Wright, unlike Lewontin, had no political stake in the issue.

Lewontin's argument has other problems, including a subtle error of statistical reasoning named Lewontin's fallacy.[27] The fallacy is to assume that the genetic differences between populations are uncorrelated with one another; if they are correlated, they become much more significant. As the geneticist A.W.F. Edwards wrote, "Most of the information that distinguishes populations is hidden in the correlation structure of the data." The 15% genetic difference between races, in other words, is not random noise but contains information about how individuals are more closely related to members of the same race than those of other races. This information is brought to light by the cluster analyses, described earlier in this chapter, which group people into populations that correspond at the highest level to the major races.

Despite the misleading political twist on Lewontin's argument, it became the centerpiece of the view that racial differences were too slight to be worth scientific attention. The assertion left the ugly implication that anyone who thought otherwise must be some kind of a racist. The subject of human race soon became too daunting for all but the most courageous and academically secure of researchers to touch.

A frequent assertion of those who seek to airbrush race out of human variation is that no distinct boundaries can be drawn between one race and another, leaving the implication that races cannot exist. "Humanity cannot be classified into discrete geographic categories with absolute boundaries," proclaims the American Association of Physical Anthropologists in its statement on race.[28] True, races are not discrete entities and have no absolute boundaries, as already discussed, but that doesn't mean they don't exist. The classification of humans into five continental based races is perfectly reasonable and is supported by genome clustering studies. In addition, classification into the three major races of African, East Asian and European is supported by the physical anthropology of human skull types and dentition.

A variation on the no distinct boundary argument is the objection that the features deemed distinctive of a particular race, like dark skin or hair type, are often inherited independently and appear in various combinations. "These facts render any attempt to establish lines of division among biological populations both arbitrary and subjective," states the American Anthropological Association's statement on race.[29] But as already noted, races are identified by clusters of traits, and to belong to a certain race, it's not necessary to possess all of the identifying traits. To take a practical example of what the anthropologists are talking about, most East Asians have the sinodont form of dentition, but not all do. Most have the EDAR-V370A allele of the EDAR gene, but not all do. Most have the dry earwax allele of the ABCC11 gene, but all not all do. Nonetheless, East Asian is a perfectly valid racial category, and most people in East Asia can be assigned to it.

Even when it is not immediately obvious what race a person

belongs to from bodily appearance, as may often be the case with people of mixed-race ancestry, race can nonetheless be distinguished at the genomic level. With the help of ancestry informative markers, as noted above, an individual can be assigned with high confidence to the appropriate continent of origin. If of admixed race, like many African Americans, each block of the genome can be assigned to forebears of African or European ancestry. At least at the level of continental populations, races can be distinguished genetically, and this is sufficient to establish that they exist.

# 6

---

# SOCIETIES AND INSTITUTIONS

It is not yet common practice to link the current social and national habitus of a nation to its so-called "history," and especially to the state-formation process it has experienced. Many people seem to have the unspoken opinion that "What happened in the twelfth, fifteenth or eighteenth centuries is past—what has it to do with me?" In reality, though, the contemporary problems of a group are crucially influenced by their earlier fortunes, by their beginningless development.

—Norbert Elias[1]

Chinese society differs profoundly from European society, and both are entirely unlike a tribal African society. How can three societies differ so greatly when their members, beneath all the differences of dress and skin color, resemble one another so closely in terms of the set of behaviors that comprise human nature? The reason is that the three societies differ greatly in their institutions, the organized patterns of behavior that structure a society, equip it to survive in its environment and enable it to compete with neighboring groups.

The institutions of Chinese, European and African societies have been deeply shaped by their respective histories as each responded to the specific challenges of its environment. The historical developments that shaped some of these institutions are described below. But it must first be noted that a society's institutions, despite their rich cultural content, are not autonomous; rather, they are rooted in basic human social behaviors. These social behaviors, as described in chapter 3, lie at the foundation of human existence as a social species. They include an instinct to cooperate vigorously with members of an in-group, to obey the in-group's rules and to punish those who deviate. There is an instinct for fairness and reciprocity, at least among members of the group. People have an intuitive morality, which is the source of instinctive knowledge that certain actions are right or wrong. People will fight to the death to protect their own group or attack that of others. Probably all these social behaviors, to one degree or another, have a genetic basis although, with the few exceptions already described, the specific underlying genes have yet to be identified.

Social institutions are a blend of genetics and culture. Each major institution is based on genetically influenced behaviors, the expression of which is shaped by culture. The human instincts for exchange and reciprocity probably undergird much of economic behavior but obviously the expressions of it, from farmers' markets to synthetic collateralized debt obligations, are cultural. "The innate mental capacities of humans underlie personal exchange. These genetic features provide the framework for exchange and are the foundation of the structure of human interaction that characterizes societies throughout history," writes the economist Douglass North, an authority on institutions.[2] The exact mix of genetic predispositions and culture in institutions has yet to be resolved, he notes.

Warfare, religion, trade and law are social institutions found throughout the world. Warfare is based on the deep-seated instinct to protect one's family and group, as well as on predatory motives, such as stealing the women or property of others. The instinct for religious behavior, found in every society, was essential for group cohesion among early human communities and continues to play a leading role in modern societies, even though other institutions have assumed many of its former roles. Trade, as noted, is founded on the human instincts for exchange and reciprocity. Law is rooted in several complex social instincts, including those for following rules, punishing violators of social norms and the sense of personal transgression that underlies self-punishment and shame.

Without knowing the nature of the genes involved in social behavior, it's impossible at present to disentangle the respective roles of culture and genetics in shaping social institutions. But language may provide a relevant example. The rules of grammar are so complex that it's hard to think every infant learns them from scratch. Rather, there must be neural machinery that both generates rules of grammar and predisposes children to learn whatever language they hear spoken around them. The role of the genes is to set up this neural learning machine. But culture provides the entire content of language.

It is notable that the cultural component of language changes surprisingly quickly: the English of 700 years ago is barely comprehensible today. The genetic machinery has presumably stayed rather constant, given that the fundamental nature of language seems the same around the world.

A similar fusion of genetics and culture is probably present in religion. The fact that every known society has a religion suggests that each inherited a propensity for religion from the ancestral human population. The alternative explanation, that each society independently invented and maintained this distinctive human behavior, seems less likely. The propensity for religion seems instinctual, rather than purely cultural, because it is so deeply set in the human mind, touching the emotional centers and appearing with such spontaneity. There is a strong evolutionary reason, moreover, that explains why religion may have become wired into the neural circuitry. A major function of religion is to provide social cohesion, a matter of particular importance among early societies. If the more cohesive societies regularly prevailed over the less cohesive, as would be likely in any military dispute, an instinct for religious behavior would have been strongly favored by natural selection. This would explain why the religious instinct is universal. But the particular form that religion takes in each society depends on culture, just as is the case with language.

The surprising longevity of many social institutions is commonly attributed to culture alone. Despite the malleability of culture and its ephemeral shifts under the influence of fashion, some cultural forms can persist for many generations and the material aspects of culture can be immensely stable—the spear has been around for millennia. But it's just as likely that in social institutions that are a blend of culture and genetics, it is the genetic component that provides the stability. Genetically based social behavior takes many generations to change whereas culture tends to drift. Even in cases where stability would seem to confer great benefit, culture can shift quite dramatically in just a few centuries. Despite the advantage of constancy in communication, languages change every generation. Religions too depend strongly on the appearance of constancy and antiquity, yet their cultural forms change quite rapidly, as is seen in the shifting shapes of Protestantism in the United States; the Puritans gave way to Congregationalism which was succeeded by Methodism, which peaked around 1850 and was overtaken by the Baptists.

The genetically based social behaviors that undergird institutions can, like any other hereditary trait, be modulated by natural selection. Human social nature is much the same from one society to another, but slight variations in social behaviors can probably generate significant and long enduring differences in a society's institutions.

A small difference in the radius of trust may underlie much of the difference between tribal and modern societies. The genetic basis of this behavior is unknown and so cannot be measured. But races and ethnicities are known to differ, for instance, in the structure of the MAO-A gene that controls aggression, as noted in chapter 3, and the differences in this gene may have been shaped by natural selection.

Institutional continuity that extends over many centuries, and over millennia in the case of China, may thus reflect the stability provided by the institutions' genetic components. One indication of such a genetic effect is that, if institutions were purely cultural, it should be easy to transfer an institution from one society to another. But American institutions do not transplant so easily to tribal societies like Iraq or Afghanistan. Conversely, the institutions of a tribal society would not work in the United States—indeed, many of them would be illegal—even if Americans could figure out what tribe they belonged to. Afghans, in order to survive in conditions where central government is usually weak, have had to rely on tribal systems for protection over the centuries, and tribal institutions require behaviors—like blood revenge and the killing of female relatives deemed to have dishonored the tribe—which differ from those that are successful in, for instance, Scandinavian democracies.

## The Great Transition

Perhaps the most dramatic example of a human society adapting through institutional change is the transition from nomadic hunter-gatherer societies to settled groups, which started only 15,000 years ago. The new institutions of settled society required a thorough makeover of human social behavior. That may explain in part why it took modern humans so long to accomplish what might seem an obviously desirable goal, that of settling in one place instead of roaming about and owning only what could be carried.

The great transition to settled life was not a single event. The ancestral human population was already dispersed across the globe when the transition started to occur. In each continent, the necessary behavioral changes occurred independently and took many generations to spread to almost everyone. Just as the populations of Europe and East Asia acquired pale skin through different genetic mechanisms, so too they independently developed the new social behaviors required to adopt a settled mode of existence. The seeds of difference between the world's great civilizations were perhaps present from the first settlements.

The institutions of hunter-gatherer societies differed greatly from those of the settled societies they became. Hunter-gatherer bands, to judge by the behavior of living hunter-gatherers, consisted of just 50 to 150 people; when they grew larger, quarrels would break out and lead to division, usually along kinship lines.

Within the hunter-gatherer groups, there were no headmen or chiefs. Strict egalitarianism prevailed and was enforced. Anyone who tried to boss others about was firmly discouraged and, if that failed, killed or ostracized. Most hunter-gatherers have no property apart from the few personal belongings that can be carried. Their economies are therefore rudimentary and do not play a major part in their survival.

Genetically, hunter-gatherer systems probably gain stability from the fact that variance is suppressed by egalitarianism. Individuals with exceptional qualities, such as great intelligence or hunting skill, cannot take direct advantage of such talents to have more children because of rules that require a catch to be shared with others. The social behavior of hunter-gatherer groups thus had no particularly strong driving force toward change.

What made settlement desirable, despite the risks, was population pressure. Hunter-gatherers require a large amount of land to provide the plants and animals they consume. After a time, even with high mortality due to incessant warfare, the available land started to run out. There was little choice but to make more intensive use of existing resources, for instance by gathering and planting wild grass seed and by controlling and penning wild animals. These practices eventually led, as much by accident as by design, to the invention of agriculture some 10,000 years ago.

The first settlements induced profound changes in human social behavior. Hierarchical systems were essential for organizing the larger numbers of people in settled communities. The egalitarianism of hunter-gatherers was abandoned. People learned the unaccustomed skills of obeying a boss. Their mental world too was transformed. The settled communities accumulated surpluses for the first time, and these could be traded. The management of these surpluses required a new kind of skill, and their defense entailed new forms of military organization.

In hunter-gatherer societies, the only division of labor was between the sexes: the men hunted and the women gathered. In settled societies, there was specialization of labor. In the wake of the specialization followed disparities of wealth.

The social and genetic variance of the society was greatly increased by these changes. A person with social skills and intelligence had a reasonable chance of getting richer, something that was seldom possible in a hunter-gatherer society, where there were no disparities and no wealth to speak of. Inequality in place of egalitarianism may not seem a good exchange, but the switch was essential for the new social structures required to operate large settled societies.

The elites that emerged in the first settled societies were able to raise more surviving children. They developed a keen interest in passing on their advantages of wealth and rank. But if the rich have more children and the population remains the same size, some children of the rich must descend in social rank. The social behaviors of the elites could thus trickle down genetically into the rest of society. The ability of the rich to produce more surviving children created for the first time a powerful mechanism whereby natural selection could enhance successful behaviors. In societies where aggression paid, aggressive men would have more children. In those in which conciliation or trading abilities carried a payoff, people with these traits would leave the larger imprint on the next generation.

The rapid adoption of new social behaviors was required for reasons both internal and external to the new societies. Within each society, people needed a very different set of skills as they adapted to new institutions like specialization of labor. And the society itself had to adapt to external pressures, such as extracting resources from a changing environment and surviving in battle against other groups. Consider how radically two critical institutions, warfare and religion, changed in nature during the transition to settlement.

Warfare is an institution doubtless inherited from the joint ancestor of chimps and humans, given that both species practice territorial-based aggression. To judge from the behavior of living hunter-gatherer societies, harsh initiation rites at puberty taught young men to bear pain without flinching. Because members of a hunter-gatherer band or tribe are usually quite highly related to one another, kinship was a strong element of group cohesion. In settled societies, kinship was abandoned as an essential basis of military cohesion once population grew beyond a certain size. Leaders took advantage of the fact that men who had become habituated to hierarchy in daily life were willing to accept military discipline.

Religion too underwent a thorough makeover in settled societies. In hunter-gatherer groups, religion is often centered around communal dances. The dances are long and vigorous and extend far into the night. There is something about rhythmic movement in unison that instills a sense of belonging to a group. There are no priests among hunter-gatherers; everyone in the group is an equal celebrant. People communicate with their gods directly, usually when some members of the group go into a trance induced by the dancing itself or by drugs.

In settled societies, by contrast, religious officials emerged as intermediaries between the people and their gods. The dancing was repressed: it represented a threat to religious authority since it allowed people to communicate with their gods directly instead of depending on the priests' interpretations. Knowledge of the gods no longer resided in songs and verbal tradition; it was assembled in religious doctrines expounded by the priests.

Religion in early societies assumed a central structural role, with the ruler often appointing himself chief priest. The pharaoh of ancient Egypt was chief priest; Roman emperors often took the title of pontifex maximus. In early settled societies that lacked any formal system of justice or the apparatus of police and courts, religion and fear of the gods' displeasure were essential means of maintaining order.[3]

People responded for the most part in cultural ways to the changing nature of institutions like warfare and religion. But both behaviors probably have an instinctual or genetic basis that can be adapted over the generations, just like any other form of behavior. In a tribal society, such as the Yanomamö of Venezuela and Brazil, aggressive men are valued as defenders in the incessant warfare between villages, and those who have killed in battle—the *unokais*—have on average 2.5 more children than men who have not killed, according to the anthropologist Napoleon Chagnon.[4] In other kinds of society, however, highly aggressive people are unlikely to prosper and will on balance have fewer children; the genetic predisposition for aggression will therefore fade over the generations, which is probably one reason why modern societies are less violent than those of medieval and earlier times.

Human social behaviors, from aggression to empathy, shape the institutions of each society, although all the details are supplied by culture. Since these institutions must change as a society's ecological and military situation changes, every aspect of human social behavior is under constant pressure from natural selection. The great transition from hunting and gathering to settled life subjected human social nature to one set of pressures. Then began another remolding process, equally extensive—that of transforming the new villagers into the subjects of empires.

# From Village to Empire

Social anthropologists usually take care not to imply that human societies have evolved, lest it seem that those that are further evolved are more advanced than others and hence superior to them. But there does seem to have been extensive evolution in human social behavior in forming civilizations such as those of ancient Egypt, Mesopotamia and China. All seem to have evolved through the same sequence of steps, or at least along parallel paths, when confronted with the same challenge.

A driving force in all these cases was demography. After the first settlements, populations started to grow. The first settlers lived in villages of perhaps 150 people. Villages would then start to cooperate, both for large agricultural projects and for defense. The larger numbers of people living in these local groups then had to be organized in some way, and this requirement led to hierarchical societies led by a chief. Warfare exerted a selective pressure under which the more cohesive societies destroyed or absorbed those that were less well organized. There was a dramatic change in human social nature that underlay a vast change in the maximum size of human societies.

Hunter-gatherer groups tend to split in two when they have more than 150 or so members; by the time of the first urban civilizations that started to arise some 5,000 years ago, people were living in cities with populations of 10,000 to 100,000.

The first chiefs secured their political power by also holding religious office. They ran their chiefdoms as family affairs, with their relatives forming a hereditary elite. But a group of chiefdoms was not a stable situation, especially if they occupied a region whose agricultural resources were circumscribed by mountains or deserts. Because of such geographical limits, the chiefdoms would impinge on one another if any tried to expand. These conditions made warfare almost inevitable.

Within each region of the world, it was war between chiefdoms that led to the emergence of the first anarchic states. "Historical or archaeological evidence of war is found in the early stages of state formations in Mesopotamia, Egypt, India, China, Japan, Greece, Rome, northern Europe, central Africa, Polynesia, Middle America, Peru and Columbia, to name only the most prominent examples," writes the anthropologist Robert Carneiro.[5]

Chiefdoms generally fought one another for territory, killing or expelling those whose lands they took over. But at a certain scale of operations, when the population was dense enough and sufficiently productive to support a ruling class, the larger chiefdoms developed into states. The states fought not just for land but also for population; instead of driving conquered people off their territory, an empire would subjugate them as part of the state's manpower.

Growing too large and complex to be managed by the ruler's family, the states developed their own cadre of officials. War between rival chiefdoms in a region could be a sanguinary affair. But once a single ruler had unified a region, there was a much greater degree of stability and order.

A general pattern in world history is that states first developed in regions of high population density, particularly along the banks of major rivers where irrigated agriculture was easy. The ancient Egyptian state began at around 3100 BC when Narmer, the ruler of the southern chiefdom of the Nile, defeated the northern chiefdom and created a unified system.

The Sumerian civilization developed around the same time along the Euphrates River in the region that is now Iraq. In India the Harappan civilization emerged in the Indus River valley. The Chinese state was built by consolidation of the settlements that arose along the Yellow River and Yangtze River valleys.

All these first generation states began to emerge some 5,000 years ago in the Old World. The process was much delayed in the New World because the population pressure necessary for state development did not begin until long after 15,000 years ago, when the first inhabitants crossed from Siberia to Alaska via Beringia, the now sunken land bridge that connected the two continents. In Mesoamerica, the Olmec state started to flourish around 1500 BC. In South America, the Moche state began around 100 AD, and the Inca empire, the most advanced state of South America, did not emerge until the 12th century AD.

The tight historical relationship between state formation and population size is evident when looking at the less easily habitable regions of the world. There are no states in the Arctic regions, sparsely inhabited by the Eskimo people. In Polynesia, there are only chiefdoms, probably because the carrying capacity of most islands permits only small populations. A major exception is Hawaii, but King Kamehameha did not unite the islands until 1811 AD.

A major region of slow population growth was Africa south of the Sahara. The continent suffers from a lack of navigable rivers, and disease makes many regions hard to inhabit. Some of Africa's chiefdoms had grown into large kingdoms, such as the Ashanti empire in Ghana, the Ethiopian empire, and the Shona kingdom in Zimbabwe by the time Europeans arrived and thwarted their further development. In 1879 a Zulu army armed with spears and oxhide shields defeated a British force armed with modern weapons at the battle of Isandlwana. But throughout much of Africa, the lack of dense populations and large scale warfare, two essential ingredients in the formation of modern states, prevented such structures from arising. Africa south of the Sahara remained largely tribal throughout the historical period, as did Australia, Polynesia and the circumpolar regions.

The evolution of human social behavior was thus different and largely or entirely independent on each continent. States had developed in the Middle East, in India and in China by around 5,000 years ago, and in Central and South America by about 1,000 years ago. For lack of good soils, favorable climate, navigable rivers and population pressure, Africa remained a continent of chiefdoms and incipient empires. In Australia, people reached the tribal level but without developing agriculture; their technology remained that of the Stone Age into modern times.

# Human Behavior in History

Although historians usually focus on states and how the actors within them seize the levers of power, in the long term it is institutions that are more important determinants of a society's fate. Being built on ingrained social behavior, institutions may endure for generations and resist even the most catastrophic events. Russians were still Russians after Stalin, Chinese remained Chinese under Mao Tse-tung; even Hitler was largely an aberration in German history.

History has little coherence when analyzed in terms of individuals or even nations. But when seen in terms of the institutions developed by different civilizations and races, the outline of a logical development emerges. Though there is still a large random element, the broad general theme of human history is that each race has developed the institutions appropriate to secure survival in its particular environment. This, then, is the most significant feature of human races: not that their members differ in physical appearance but that their society's institutions differ because of slight differences in social behavior.

A landmark analysis of human history in terms of social institutions has recently been written by the political scientist Francis Fukuyama. His thesis, describing how each of the major civilizations adapted its institutions to its local geography and historical circumstances, provides a roadmap of human social adaptation and the different paths taken by each civilization.

Fukuyama's premise, like that of North quoted above, is that institutions are rooted in human social behavior. "The recovery of human nature by modern biology . . . is extremely important as a foundation for any theory of political development, because it provides us with the basic building blocks by which we can understand the later evolution of human institutions," he writes.[6]

A pillar of human social nature is the tendency to favor family and close kin, and this is the root of tribalism. Tribal societies were the first form of human political organization, given that the hunter-gatherer bands in which humans have passed most of their existence were probably organized as tribes from an early date. A tribe consisted of bands that exchanged women in marriage. Tribal organization is highly flexible, and tribes can grow to vast sizes capable of considerable undertakings: the Mongols, whose empire stretched from the Pacific Ocean to the borders of Europe, were tribally organized. The weak point of the tribal system is succession: when a strong leader dies, the chiefs of the component lineages tend to fight one another to succeed him, and the whole coalition may break down into smaller, feuding entities, as was the fate of the Mongol empire.

Tribes are organized on the basis of lineages traced through the male line of descent. Within a tribe, two lineages may fight each other or join together to vie with a third. Because all lineages descend from a founding patriarch, any two lineages can find a joint ancestor to prove their kinship and affinity as allies. Anthropologists call tribes segmentary societies because of the way in which the different lineages or segments can be fitted together for particular social purposes.

The tribal system is so strong, in Fukuyama's view, that even in most modern states it never fully disappears. Rather, the state apparatus is layered on top and is in constant tension with it. In China, officials use their positions to advance their kindreds' interests, regardless of the state's. The problem is as pertinent in China today as at any point in the past. Even in Europe and the United States, where family relationships are less intricate and tribes no longer exist, nepotism is far from unknown.

Tribalism is the default state of early human societies, just as autocracy is the default state of modern ones. Tribal societies have existed probably from the beginning of the human species, and many still exist in the present day. The inhabitants of Spain, France, Germany and England were tribal peoples before and after their conquest by the Roman state. In China, tribal chiefdoms did not start to disappear until the 4th century BC; in much of Africa and the Middle East, tribal organization remains a potent force.

Given the pervasiveness of tribalism, how did modern states ever get started? Fukuyama's approach to the answer is to consider the differences between modern states in order to understand which of their features are the most significant. Surveying the modern states that arose in China, Europe, India and the Muslim world, he finds that all had to confront the same principal challenge, that of suppressing tribalism so that the state's authority could prevail, but that each accomplished this goal in very different ways.

China achieved a modern state a millennium before Europe. This precocious advance may have had a lot to do with the nature of the plain between the Yangtze and Yellow rivers. The territory is well suited both for agriculture, which leads to population growth, and for warfare, the two principal propellants of state formation. A relentless process of consolidation forced tribal systems to yield to states.

In 2000 BC, a large number of political entities—traditionally put at 10,000—existed in the Yellow River valley. By the time of the Shang dynasty in 1500 BC, these had dwindled to some 3,000 tribal chiefdoms. The Eastern Zhou dynasty began in 771 BC with 1,800 chiefdoms and ended with 14 entities that were much closer to states. During the ensuing Warring States period, which lasted from around 475 BC to 221 BC, the 7 remaining states were reduced to 1.

China became unified in 221 BC when the state of Qin managed to defeat its six rivals during the Warring States period. This was the culmination of some 1,800 years of almost incessant strife, during which the demands of warfare shaped the distinctive lineaments of the Chinese state.

The tribal system endured as long as the Yellow River valley was relatively uninhabited. A weaker tribe could just move elsewhere. As population density increased, the choice became to fight or to be extinguished.

The pressure on the tribes arose through their mode of fighting. Being based on male lineages, the fighting was done by nobles in chariots, with each chariot requiring a logistics train of some 70 soldiers. With incessant wars, the number of available nobles was eventually depleted. In desperation, some chiefdoms during the Zhou period developed an alternative mode of warfare, that of impressing the peasantry into infantry armies.

This was not a simple transformation, given that it did not recommend itself to either the nobility or the peasantry. Moreover, it required a complex and imaginative set of institutional changes. Larger armies required raising more taxes to support them. Extracting taxes from the population made necessary a class of officials loyal to the state, not to particular tribes.

These changes began in several states, but it was in Qin, the most westerly of the seven warring states, that the reforms were pushed furthest. "Groundwork for the first truly modern state was laid in the western polity of Qin under Duke Xiao and his minister Shang Yang," Fukuyama writes.[7]

The Qin leaders built a modern state because they recognized explicitly that the noble lineages of the tribal system were an impediment to the state's power. Shang Yang abolished the hereditary offices held by the nobles in favor of a 20 rank system based on military merit. This change meant that all officeholders now owed their position and loyalty to the state, not to their tribe or lineage.

Not only was the bureaucracy appointed on merit, it was also rewarded on performance. Important items such as land, servants, concubines and clothing were distributed to those who served the state well.

In a second bold stroke of social engineering, Shang Yang let peasants own land directly instead of having to work fields owned by the nobility. The peasants were now directly beholden to the state and owed their taxes to the state, not the nobility.

But this was no agricultural reform designed for the peasants' benefit. Previously the peasants had worked under the nobles' supervision. Shang Yang had them reorganized into groups of five or ten households, which were required to supervise one another and report crimes to the state. Failure to report was punishable by death.

"If the people are stronger than the government, the state is weak; if the government is stronger than the people, the army is strong," states the treatise attributed to Shang Yang.[8] This was the point of the whole exercise. The peasantry was controlled and taxed. The bureaucrats administered the state and raised the taxes to finance a mass peasant army.

The westerly Qin state, though long regarded as something of a backwater, now had the political organization required to pay for a substantial army. With this force, the Qin king was able in 221 BC to defeat his six rivals and unify China. Unification brought to an end the deadly game of the 254-year Warring States period, during which 468 wars were fought between the rival players.

The Chinese had invented the modern state more than a thousand years before Europe did. A finishing touch was added when the Mandarin examination system was instituted in 124 BC under emperor Wu. Besides an army, tax collection systems, registration of the population and draconian punishments, China had another institution, one that the sociologist Max Weber considered the defining mark of a modern state, that of an impersonal bureaucracy chosen by merit.

The Chinese state arose because tribal organization could not handle the demands of the Chinese style of warfare. With China as a template, comparisons can be made with how other civilizations developed modern states. Europe, for instance, after the disintegration of the Roman empire, had a period analogous to the Eastern Zhou dynasty when its tribes were developing into states, symbolized by the process in which the king of the Franks became instead the king of France. During this period, the number of European polities was reduced from about 500 to 25. But Europe then deviated from the Chinese pattern, because this process of reduction was not followed by a final unification, a Warring States epoch in which one state emerged the winner.

Why did no counterpart of the state of Qin arise to conquer all of Europe? One reason may be that state building came a thousand years later to Europe, by which time feudalism had secured a stronger foothold than in China. The local chieftains could not be dispossessed in the Shang Yang style. Kings had to negotiate with them. So no European state became strong enough to dominate all the others in any sustained way; after the Romans, attempts at empire in Europe were always partial and short-lived.

Another reason is that barriers of geography and culture were more formidable in Europe than in the Yellow River valley. Europe is divided by mountain ranges and rivers, and within these natural compartments emerged differences of religion and language. These impediments made it far harder to construct a unified European state.

China was able to develop the institutions of an autocratic state, ones so effective that China for most of its modern history has been unified, though punctuated by short, disruptive periods of disunity. Despite its autocratic nature, the state was several times conquered by one or another of the various tribal pastoralists, like the Mongols or the Manchu, who roamed the steppes beyond China's northern border. Yet these conquerors found that to rule China they had to abandon their tribal ways and adopt Chinese institutions.

A striking counterpoint to the Chinese pattern of development is provided by India. By the 6th century BC, the first states had developed in India, as in China. But whereas in China there followed 500 years of incessant warfare, India did not undergo such a process, perhaps because population was less dense. The principal shaper of Indian society was not war but religion. Brahmanism divided society into four classes, those of priests, warriors, merchants and everyone else. The four classes were subdivided into hundreds of endogamous occupational castes. This system, layered on top of the tribal divisions, proved so strong that no government could overrule it. India thus created a strong society and a weak state, the inverse of the Chinese situation in which, then as now, the people have seldom challenged the state-controlled institutions.

The state was so weak, in fact, that it has seldom been unified. The Maurya empire ruled all but southern India for a century after

321 BC but, unlike the Qin in China, did not seek to impose its own institutions throughout the empire. When it disintegrated, it was only foreign invaders who showed an interest in integrating the subcontinent, such as the Mughals and then the British.

There is no basis in Indian political institutions for a tyrannical state, whereas in China since the Qin, the state has always assumed the right to tell its citizens what to do. Yet China, for all its precocious modernity, never developed the rule of law, the concept that the ruler should be subject to some independent body of rules. In India, Fukuyama writes, law "did not spring from political authority as it did in China; it came from a source independent of and superior to the political ruler."

India did not develop the formal mechanisms contrived by European states for holding the ruler accountable to the law. But from the earliest times, the religious law was a central institution that circumscribed the power of the state. The respective institutions developed by India and China had a major role in shaping their different histories up until the present day. From the Great Wall to the Three Gorges Dam, the Chinese state has never hesitated to force costly public works on its citizens, who have no way to object or resist. In India, by contrast, the government cannot propose a new airport or factory site without facing vociferous public protest.

In China, tribalism was suppressed by direct actions of the state; in India, by religion. The most inventive way of undermining tribalism was developed in the Islamic world by the Abbasid dynasty and brought to perfection by the Ottomans. This was the institution of military slavery, in which both the military and the bureaucratic elite of the empire were made up of slaves. A matter of possible envy to any chief executive who has tried to impose his wishes on a resistant bureaucracy, the sultan could order the execution of any slave bureaucrat, from those of the lowest ranks up to the grand vizier. The slaves, at least in principle, were forbidden to have families or, if allowed to marry, their sons were prohibited from becoming soldiers or succeeding their fathers in office.

The strong human instinct of favoring one's family or kin was thus thwarted. The elite slave officeholders were the empire's aristocracy for just a generation; their children had to join the general population. As for the problem of where to acquire the slaves (Islam forbids the enslavement of Muslims), the Ottomans addressed that with the *devshirme*, a system in which talent spotters would visit Christian provinces, notably Serbia, and require local priests to provide a register of all boys baptized in the region. The most promising were abducted from their parents, whom they would never see again. They were then converted to Islam and trained either to become senior administrators or to join the Janissaries, an elite military group.

As bizarre and inhuman as the institution of military slavery may seem, it shows the lengths to which a state would go in order to thwart tribalism and secure a caste of administrators responsive to the ruler's commands. The institution was invented by the Abbasids, the Islamic dynasty that held sway in the Near East from 750 to 1258 AD, but they found it impossible to rule a far-flung empire with Arab tribal organization. It was further developed by the Ayyubid Sultans of Egypt, who created an army, the Mamluks, from slaves captured from Turkic tribes and the Caucasus. In slave armies with no nobility or kinship system, people could be promoted purely on merit. This and the commanders' exclusive loyalty to their sultan were key to the success of the Mamluk and Janissary armies.

The Mamluks of Egypt saved the Islamic world by defeating an invading Mongol army in 1260 at the battle of 'Ain Jalut. But the commander of the Mamluk army, Baibars, then overthrew his master and became sultan of Egypt. The Mamluks continued to be a formidable military force for several decades, defeating three more Mongol invasions. But wealthy Mamluks then found ways to defeat the prohibition on bequeathing wealth to their descendants, and the system gradually retribalized itself. "The one-generation nobility principle worked against the basic imperatives of human biology, just as the impersonal Chinese examination system did," Fukuyama observes. "Each Mamluk sought to protect the social position of his family and descendants."

The Mamluk system started to decay. Ridden by factionalism and handicapped by their disdain for the new firearms now dominating battlefields, the Mamluks were defeated by the Ottomans of Turkey in 1517.

Military slavery served the Ottomans well and for a time made possible the continual conquests on which the economy of the state depended. But when the Ottoman expansion ceased, the Sultans first allowed the Janissaries to marry and have children and then permitted their sons to enter military service. This made the *devshirme* unnecessary. It also destroyed the basic purpose of the system, that of preventing the emergence of a hereditary elite. The institution began to decay, and the slow disintegration of the Ottoman state proceeded.

The fourth major civilization on the Eurasian continent, that of Europe, developed a complex set of institutions that is more easily understood in comparison with the somewhat simpler cases of China, India and the Islamic world. A distinctive feature of European states is that, having escaped from tribalism, they then developed an institution contrived by none of the other three civilizations—that of a means for society to control a strong leader.

There emerged in Europe the concept of the rule of law, a consensus among society and the elite that the ruler was not sovereign, the law was sovereign. Second, Europe and particularly England developed ways to hold the king accountable to the law. This structure had the virtue of allowing the ruler to be strong but subject to institutional restraint.

The Chinese state, from the days of the Qin, was an efficient, bureaucratized autocracy. Yet to this day, China has never developed the rule of law. Its emperors, and now the Chinese politburo, make the law but are not accountable to it and do not have to obey it themselves. China can always force its people to build a Great Wall or its equivalent. But its great disadvantage as a strong state is to be defenseless against a bad emperor, of whom the most recent was Mao Tse-tung.

A central part in the development of European institutions was played by religion. Religion, Fukuyama argues, was critical first in detribalization and second in instituting the rule of law. The essence of the tribal lineage was the descent of property through the male line. But producing a male heir under medieval conditions of short life expectancy and high infant mortality was far from a sure thing. So the tribes had various strategies for keeping wealth within the lineage. These included cousin marriage, divorce if a woman bore no heir, adoption and the levirate (marrying of widows to their husband's brother). In addition, women were not allowed to own property.

The church opposed all these heirship strategies, not because of anything in existing Christian doctrine but because it had a better idea: that people should leave their property to the church instead of their heirs. By the end of the 7th century, a third of the productive land in France belonged to the church. The tribes of Europe, whether Frankish, Anglo-Saxon, Slavic, Norse or Magyar, found that conversion to Christianity soon separated them from their property, robbed them of their influence and paved the way to feudalism.

In the fragmented political conditions of medieval Europe, the church became rich and powerful but started to develop tribal or nepotistic problems of its own. Its priests became keenly interested in passing on their property and offices to their kin. Pope Gregory VII forced priests to become celibate so that their loyalty would be to the church and not to their kin.

Gregory was also at the center of the historic confrontation between the pope and the Holy Roman Emperor over investiture, the question of who had the right to appoint bishops. Gregory excommunicated Henry IV, who in turn tried to depose Gregory. But the church prevailed, forcing Henry to come to the pope's residence at Canossa in 1077 and wait barefoot in the snow for three days to receive Gregory's absolution.

The church used its power to back the idea of law, first of the Justinian code, a Byzantine codification of Roman law that was rediscovered around 1070 AD, and then of canon law, the synthesis by Gratian of church laws through the centuries. Because law had the church's authority, sanctioned by a higher power, there emerged in Europe the novel idea that the ruler could not rule in defiance of the law and indeed owed his position to his role in upholding the law.

Feudal Europe was a collection of local barons installed in largely impregnable castles. Kings tended to be the first among equals and had to negotiate with others to exercise power. They were obliged to take account of the concept that the law and not the king was sovereign. They could not tax or conscript peasants because those rights belonged to the feudal lords. Nor could they seize land because of the property rights conferred by the feudal system.

National states emerged in Europe as part of a struggle between the king, the elites and other sources of power. The kings were seldom absolute rulers. The limitation on their powers was taken furthest in England, where Parliament raised its own armies, executed Charles I and forced James II to abdicate. The English state thus constructed a system, later followed by other European countries, in which the ruler was subject to the law and in which a representative body held him accountable to it.

"Once this package had been put together the first time," Fukuyama writes, "it produced a state so powerful, legitimate, and favorable to economic growth that it became a model to be applied throughout the world."[9]

# Effects on Social and Individual Behavior

From the broad sweep of Fukuyama's analysis, a clear pattern emerges. Each of the major civilizations of Eurasia developed a characteristic set of institutions in response to its local circumstances and history. Given that institutions rest on a basis of human social behavior, and that culture feeds back into the genome, it is plausible to suppose that Chinese, Indians, Ottomans and Europeans have all adapted over the generations to their particular social conditions. Because of this evolutionary process, the four civilizations remain distinct today.

The social institutions of the four civilizations had considerable inertia, meaning that they changed very slowly over time. Institutions that endure for many generations are strong candidates for being rooted in a genetically framed social behavior that maintains their stability. East Asian societies have a distinctive character, tending to be efficient autocracies. Singapore, for instance, endowed at independence with a cultural heritage of English political institutions, nevertheless has become a looser version of the autocratic Chinese state despite retaining the outward forms of a European one.

A similar continuity in social behavior is evident in Africa, which has consisted largely of tribally organized societies both before and after the episode of colonial rule. European powers prepared their colonies for independence by imposing their own political institutions. But these had been developed over many centuries for the European environment. Considering the long historical process by which Europeans had rid themselves of tribalism, it is hardly surprising that African states did not become detribalized overnight. They reverted to the kind of social system to which Africans had become adapted during the previous centuries.

In tribal systems, people very rationally look to their relatives and tribal groupings for support, not to a central government, whose usual function has been to exact taxes or military service while giving back little in return. European or American institutions cannot easily be exported to tribal societies like those of Iraq or Afghanistan because they presuppose a large measure of trust toward non-kin and are designed to operate in the public interest, not to empower the officeholder and his tribe.

Variations in human social behavior and in the institutions that embody it have far-reaching consequences. Developmental economists long ago learned that it is not just lack of capital or resources that keeps countries poor. Billions of dollars' worth of aid have been poured into Africa in the past half century with little impact on the standard of living. Countries like Iraq are rich in oil, but their citizens are poor. And countries with no resources, like Singapore, are rich.

What makes societies rich or poor is to a great extent their human capital—including the nature of the people, their levels of training, the cohesiveness of their societies, and the institutions with which they are organized. As Fukuyama notes, "Poor countries are poor not because they lack resources but because they lack effective political institutions."[10]

The same conclusion is reached in the recent book *Why Nations Fail* by the economist Daron Acemoğlu and the political scientist James Robinson. "Nations fail economically because of extractive institutions," they write, meaning institutions that allow a corrupt elite to exclude others from participating in an economy.[11] Conversely, they say, "Rich nations are rich largely because they managed to develop inclusive institutions at some point during the last 300 years."[12]

Acemoğlu and Robinson's theory is discussed further in the next chapter. Of relevance for the moment is that they and Fukuyama have independently concluded that institutions are central to the success and failure of human societies. Less clear is why institutions differ from one society to another. These differences became most evident during the profound shift in the structure of human societies that culminated in the Industrial Revolution.

There have been two major steps in the evolution of human societies, both accompanied by changes in human social behavior. The first was the transition from the hunter-gatherer existence to that of settled societies. The settled societies developed agriculture but then stagnated for hundreds of generations in what is known as the Malthusian trap: each increase in productivity was followed by a growth in population, which ate up the surplus and brought the population back to the edge of starvation. The trap could not be escaped until human social nature had undergone a second major transition. Following is the case for thinking that a deep genetic change in social behavior underlay the escape from the Malthusian trap and the transition from an agrarian to a modern society.

# THE RECASTING OF
# HUMAN NATURE

We need to confront the most blatant fact that has persisted
across centuries of social history—vast differences in productivity among peoples and the economic and other consequences of such differences.

—Thomas Sowell[1]

Each of the major civilizations has developed the institutions
appropriate for its circumstances and survival. But these institutions, though heavily imbued with cultural traditions, rest
on a bedrock of genetically shaped human behavior. And when a
civilization produces a distinctive set of institutions that endures for
many generations, that is the sign of a supporting suite of variations
in the genes that influence human social behavior.

Historians sometimes speak of national character, but although
many might agree that the German and Japanese characters, say, differ in ways that have deeply affected their respective histories, there
is less agreement on how to define the significant elements of national

character. And without some objective measure, attempts to describe national character easily slip into caricature.

How could any objective measure be found of how human nature changes over time? Surprisingly enough, such measures exist, even though they are indirect. They come from the work of economic historians, such as Maristella Botticini and Zvi Eckstein, who have documented the role of education in Jewish history, and Gregory Clark, who has reconstructed English economic and social behavior in the 600 years that preceded the Industrial Revolution.

The change brought about by the Industrial Revolution was not a visible alteration of people's lifestyles, such as living in houses instead of the wild, but a quantum leap in society's productivity. After stagnating for at least the five and a half centuries that can be

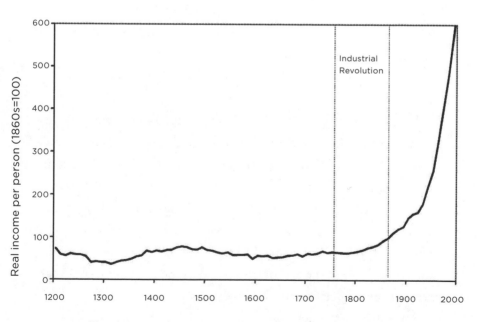

Figure 7.1. Real income per person in England, 1200–2000s

FROM CLARK, *FAREWELL TO ALMS*

documented, English wages started to soar in the mid-18th century, reflecting an astonishing increase in the level of productive work.

Productivity increases may seem like something that could excite only an economist, but it made all the difference to people's lives. Before the Industrial Revolution, almost everyone but the nobility lived a notch or two above starvation. This subsistence-level existence was a characteristic of agrarian economies, probably from the time that agriculture was first invented.

The reason was not lack of inventiveness: England of 1800 possessed sailing ships, firearms, printing presses and whole suites of technologies undreamed of by hunter-gatherers. But these technologies did not translate into better living standards for the average person. The reason was a catch-22 of agrarian economies that dates back probably to the beginning of agriculture.

The catch is called the Malthusian trap because it was described by the Reverend Thomas Malthus in 1798 in his *Essay on the Principle of Population*. Each time productivity improved and food became more plentiful, more infants survived to maturity and the extra mouths ate up the surplus. Within a generation, everyone was back to living just above starvation level.

This lack of progress has been documented by the economic historian Gregory Clark of the University of California, Davis. Because of the existence of copious historical information in England, a country untouched by hostile invasion since 1066 (William of Orange's invasion in 1688 was by invitation), Clark has been able to reconstruct many economic data series such as the real day wages of English farm laborers from 1200 to 1800. Wages were almost exactly the same at the end of the period as they had been 600 years earlier. They sufficed to buy a meager diet.

But wages were not constant throughout the period. Between 1350 and 1450 they more than doubled. The cause was not some

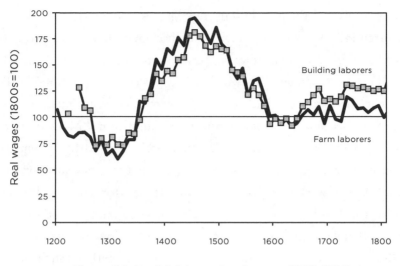

Figure 7.2. English laborers' real wages, 1200–1800.

From Clark, *Farewell to Alms*

miraculous increase in productivity—it was the Black Death, which carried off some 50% of the population of Europe. In a Malthusian-trap world, plagues are a blessing, at least for the survivors. With fewer mouths to feed, everyone ate better, and with a new scarcity of labor, workers could enjoy better wages. This era of plenty lasted a century until rising population numbers closed the jaws of the Malthusian trap once again.

In almost all societies since the invention of agriculture most people, aside from the ruling elite, have lived under these harsh conditions. England probably did not differ from other agrarian societies in Europe and East Asia between 1200 and 1800 except that the economic conditions of its Malthusian trap are unusually well documented.

Malthus, strangely enough, wrote his essay at the very moment when England, shortly followed by other European countries, was about to escape the Malthusian trap. The escape consisted of such a substantial increase in production efficiency that extra workers enhanced incomes instead of constraining them.

This development, known as the Industrial Revolution, is the salient event in economic history, yet economic historians say they have reached no agreement on how to account for it. "Much of modern social science originated in efforts by late nineteenth and twentieth century Europeans to understand what made the economic development path of western Europe unique; yet these efforts have yielded no consensus," writes the historian Kenneth Pomeranz.[2] Some experts argue that demography was the real driver: Europeans escaped the Malthusian trap by restraining fertility through methods such as late marriage. Others cite institutional changes, such as the beginnings of modern English democracy, secure property rights, the development of competitive markets, or patents that stimulated invention. Yet others point to the growth of knowledge starting from the Enlightenment of the 17th and 18th centuries or the easy availability of capital.

This plethora of explanations and the fact that none of them is satisfying to all experts point strongly to the need for an entirely new category of explanation. Clark has provided one by daring to look at a plausible yet unexamined possibility—that productivity increased because the nature of the people had changed.

Clark's proposal is a challenge to conventional thinking because economists tend to treat people everywhere as identical. None would suggest that the Stone Age economies in which New Guinean societies were living when discovered by Europeans had anything to do with the nature of New Guineans. Provided with the same incentives, resources and knowledge base, New Guineans would develop economies similar to that of Europeans, most economists would say.

A few economists have recognized the implausibility of this position and have begun to ask if the nature of the humble human units that produce and consume all of an economy's goods and services might possibly have some bearing on its performance. They have discussed human quality, but by this they usually mean just education and training. Others have suggested that culture might explain why some economies perform very differently from others, but without specifying what aspects of culture they have in mind. None has dared say that culture might include an evolutionary change in behavior, though nor do they explicitly exclude this possibility.

To appreciate the background of Clark's idea, one has to return to Malthus. Malthus's essay had a profound effect on Charles Darwin. It was from Malthus that Darwin derived the principle of natural selection, the central mechanism in his theory of evolution. If people were struggling on the edge of starvation, competing to survive, then the slightest advantage would be decisive, Darwin realized, and the owner would bequeath that advantage to his children. These children and their offspring would thrive while others perished.

"In October 1838, that is, fifteen months after I had begun my systematic inquiry," Darwin wrote in his autobiography, "I happened to read for amusement Malthus on Population, and being well prepared to appreciate the struggle for existence which everywhere goes on from long-continued observation of the habits of animals and plants, it at once struck me that under these circumstances favorable variations would tend to be preserved, and unfavorable ones to be destroyed. The results of this would be the formation of a new species. Here then I had at last got a theory by which to work."

Given the correctness of Darwin's theory, there is no reason to doubt that natural selection was working on the very English population that provided the evidence for it. The critical issue, then, is that of just what traits were being selected for.

As it happens, Clark has documented four behaviors that steadily changed in the English population between 1200 and 1800, as well as a plausible mechanism of change. The four behaviors are those of interpersonal violence, literacy, the propensity to save and the propensity to work.

Homicide rates for males, for instance, declined from 0.3 per thousand in 1200 to 0.1 in 1600 and to about a tenth of this in 1800.[3] Even from the beginning of this period, the level of personal violence was well below that of modern hunter-gatherer societies. Rates of 15 murders per 1,000 men have been recorded for the Aché people of Paraguay.

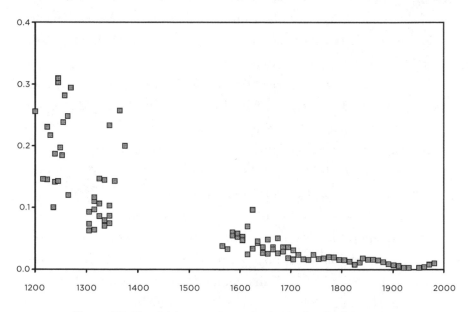

Figure 7.3. Homicide rates for males in England, 1190–2000.

From Clark, *Farewell to Alms*

Literacy rates can be estimated from the proportion of people who spell out their names on documents, such as marriage registers and court documents, instead of signing with an X. The literacy rate among English men climbed steadily from about 30% in 1580 to above 60% by 1800. Literacy among English women started from a

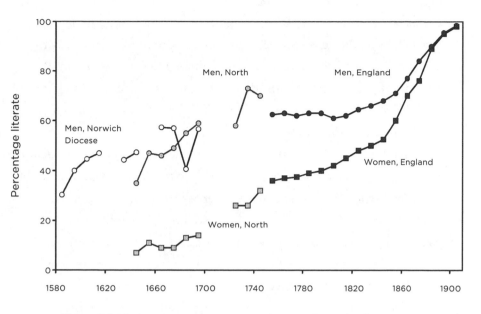

Figure 7.4. Literary rates among men and women in England, 1580-1920.

FROM CLARK, *FAREWELL TO ALMS*

lower base—about 10% in 1650—but had equaled that of men by 1875.[4]

Work hours steadily increased throughout the period, and interest rates fell. When inflation and risk are subtracted, an interest rate reflects the compensation that a person will demand to postpone immediate gratification by postponing consumption of a good from now until a future date. Economists call this attitude time preference, and psychologists call it delayed gratification. Children, who are generally not so good at delaying gratification, are said to have a high time preference. In his celebrated marshmallow test, the psychologist Walter Mischel tested young children as to their preference for receiving one marshmallow now or two in fifteen minutes. This simple decision turned out to have far-reaching consequences: those able to hold out for the larger reward had higher SAT scores and social competence in later life. Children have a very high time preference which falls as they grow older and develop more self-control. American six-year-olds, for instance, have a time preference of about 3% per day, or 150% per month; this is the extra reward they must be offered to delay instant gratification. Time preferences are also high among hunter-gatherers.

Interest rates, which reflect a society's time preferences, have been very high—about 10%—from the earliest historical times and for all societies before 1400 AD for which there is data. Interest rates then entered a period of steady decline, reaching about 3% by 1850. Because inflation and other pressures on interest rates were largely absent, Clark argues, the falling interest rates indicate that people were becoming less impulsive, more patient and more willing to save.

These behavioral changes in the English population between 1200 and 1800 were of pivotal economic importance. They gradually transformed a violent and undisciplined peasant population into an efficient and productive workforce. Turning up punctually for work every day and enduring eight hours or more of repetitive labor is far from being a natural human behavior. Hunter-gatherers do not willingly embrace such occupations, but agrarian societies from their beginning demanded the discipline to labor in the fields and to plant and harvest at the correct times. Disciplined behaviors were probably gradually evolving within the agrarian English population for many centuries before 1200, the point at which they can be documented.

Growth in productive efficiency makes all the difference to economic output, on which a population's prosperity and survival depend. In 1760, just as the Industrial Revolution was about to take off, 18 hours of labor were required to transform a pound of cotton into cloth. A century later, only 1.5 hours were needed.[5]

Better technology played a large role in the growth in efficiency. The difference was made not so much by the major inventions beloved of historians, like Richard Arkwright's water frame or James

Hargreaves's spinning jenny, but by a continuous stream of incremental improvements as workers drew from and improved upon an expanding pool of common technical knowledge.

Clark has uncovered the simple genetic mechanism through which the Malthusian economy wrought these changes on the English population: the rich had more surviving children than did the poor. From a study of wills made between 1585 and 1638, he finds that will makers with £9 or less to leave their heirs had, on average, just under two children. The number of heirs rose steadily with assets, such that men with more than £1,000 in their gift, who formed the wealthiest asset class, left just over four children.

The English population was fairly stable in size from 1200 to 1760. In this context, the fact that the rich were having more children

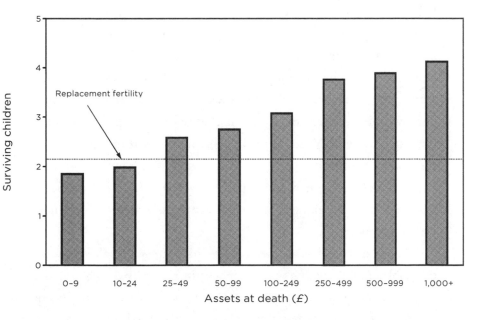

Figure 7.5. Surviving children by assets of the testator.

From Clark, *Farewell to Alms*

than the poor led to the interesting phenomenon of unremitting social descent. Most children of the rich had to sink in the social scale, given that there were too many of them to remain in the upper class.

Their social descent had the far-reaching genetic consequence that they carried with them inheritance for the same behaviors that had made their parents rich. The values of the upper middle class—nonviolence, literacy, thrift and patience—were thus infused into lower economic classes and throughout society. Generation after generation, they gradually became the values of the society as a whole. This explains the steady decrease in violence and increase in literacy that Clark has documented for the English population. Moreover, the behaviors emerged gradually over several centuries, a time course more typical of an evolutionary change than a cultural change.

That a profound change in human social behavior should evolve in just a few centuries may seem surprising, but it is perfectly possible in light of the experiments conducted by Dmitriy Belyaev on domestication (his experiments on breeding both tamer and fiercer rats were mentioned in chapter 3). Belyaev was a Soviet scientist who believed in evolution despite the anti-genetic views of Trofim Lysenko, which were then official doctrine in the Soviet Union. In a remote institute in Novosibirsk, he began to test his theory that ancient farmers had domesticated wild animals by a single criterion, that of tameness. All the many other traits that distinguish domestic animals from their wild forebears—thinner skulls, patches of white hair, floppy ears—had been dragged along in the wake of the selection for tameness, Belyaev supposed.

He began by selecting silver foxes for tameness, taking the remarkable gamble that he would see within his lifetime a change that may have taken ancient farmers many hundreds of years to accomplish. Yet within eight generations, Belyaev had bred silver foxes that would tolerate human presence. Just 40 years after the experiment had started and with 30 to 35 generations of breeding, the foxes were as tame and biddable as a dog. And just as Belyaev had predicted, the tamer foxes had white patches on their coats and droopy ears, even though these traits had not been selected for.[6]

Belyaev's work, which did not become known outside the Soviet Union until 1999, demonstrated how quickly a profound evolutionary change in behavior could occur. Assuming 25 years per generation, there would have been 24 human generations between 1200 and 1800, plenty of time for a significant change in social behavior if the pressure of natural selection were sufficiently intense.

In a broader sense, these changes in behavior were just some of many that occurred as the English population adapted to a market economy. Markets required prices and symbols and rewarded literacy, numeracy and those who could think in symbolic ways. "The characteristics of the population were changing through Darwinian selection," Clark writes. "England found itself in the vanguard because of its long, peaceful history stretching back to at least 1200 and probably long before. Middle-class culture spread throughout the society through biological mechanisms."[7]

Economic historians tend to see the Industrial Revolution as a relatively sudden event and their task as being to uncover the historical conditions that precipitated this immense transformation of economic life. But profound events are likely to have profound causes. The Industrial Revolution was caused not by events of the previous century but by changes in human economic behavior that had been slowly evolving in agrarian societies for the previous 10,000 years.

This, of course, explains why the practices of the Industrial Revolution were adopted so easily by other European countries, the United States and East Asia, all of whose populations had been living in agrarian economies and evolving for thousands of years under the same harsh constraints of the Malthusian regime. No single resource or institutional change—the usual suspects in most theories of the Industrial Revolution—is likely to have become effective in all these countries around 1760, and indeed none did.

That leaves the questions of why the Industrial Revolution was perceived as sudden and why it emerged first in England instead of in any of the many other countries where conditions were ripe. Clark's answer to both these questions lies in the sudden growth spurt in the English population, which tripled between 1770 and 1860. It was this alarming spurt that led Malthus to write his foreboding essay on population.

But contrary to Malthus's gloomy prediction of a population crash induced by vice and famine, which would have been true at any earlier stage of history, incomes on this occasion rose, heralding the first escape of an economy from the Malthusian trap. Incomes grew because the production efficiency of the English economy had been steadily increasing since 1600. It had reached such a level that, combined with the sudden rise in population, the output of the English economy became visibly larger. English workmen contributed to this spurt, Clark drily notes, as much by their labors in the bedroom as on the factory floor.

The rise in population that made England's exit from the Malthusian trap so visible was an unrelated event, in Clark's view. It had no part in causing the escape but merely amplified a process that was already under way. Clark attributes it to women's perception that the once substantial risks of death in childbirth had dropped substantially since the 17th century. In 1650 a woman who had the average number of children had a 10% chance of dying in childbirth. This formidable risk had dropped to just over 4% by the early 19th century. In 1650, 20% of women never married, and the perceived risk of doing so would have been a rational deterrent. By the early 18th century the proportion of spinsters had fallen to 10%. This and the trend to younger marriages propelled a 40% increase in English fertility between 1650 and 1800.[8]

Clark's thesis departs considerably from the mainstream views of economic historians and political economists, most of whom look to institutions to explain major issues such as world poverty and the Industrial Revolution, even though each has a different favorite, whether intellectual property rights, the rule of law or parliamentary democracy. Clark dismisses this whole category of explanations as insufficient. Many early societies, he says, had all the preconditions for economic growth that any World Bank economist could wish for, yet none did so. "Economic historians," Clark writes, "thus inhabit a strange netherworld. Their days are devoted to proving a vision of progress that all serious empirical studies in the field contradict." They are thus "trapped in this ever-tightening intellectual death spiral."

Clark's book was widely noticed and, unsurprisingly given its heterodoxy, many reviews were critical. Some reviewers dismissed Clark's thesis as peremptorily as he had dismissed theirs. Several disagreed with his assertion that England prior to the Industrial Revolution was under a true Malthusian regime, an issue of contention among economic historians. Others disputed Clark's calculation about human wealth before agriculture, which has to be inferred from that of living hunter-gatherers. However the strictly economic issues may be resolved, there were relatively few attacks on Clark's proposed mechanism of evolutionary change, the ability of the rich to leave more surviving children who would spread their genes and behavior through the population as some of them descended in social rank.

Clark has since corroborated this mechanism by devising an independent way to check it, based on the prevalence of surnames. Surnames, being passed from father to son, are effectively propagated like the Y chromosome. They track male genes, provided that wives are faithful and no one is adopted, but cases of nonpaternity and adoption were both rare in medieval England. Clark chose two sets of rare surnames, such as Banbricke, Cheveney, Reddyforde, Spatchet and Tokelove, from English records of 1560–1640. One set belonged to men rich enough to leave a will, the other to people indicted in Essex courts for burglary, poaching and crimes of violence, and therefore assumed to be among the poorest.

For rare surnames, a large fraction of the holders will typically be related. Clark found that his rich families survived through the generations much better than the poor ones. By 1851 only 8% of the richest surnames from the 1560–1640 period had disappeared, but 21% of the indicted surnames no longer existed. The poor have a greater risk of being erased from the gene pool.

But it is not the case, Clark found, that a permanent rich elite survives in perpetuity. Rather, there has been considerable social mobility in English society. Many of the rare surnames that belonged to rich families in 1560–1640 belonged to people in middle- or lower-income occupations, and some of the indicted surnames from the earlier period had risen into the gentry by 1851.

"The surname evidence confirms a permanent selection in preindustrial England for the genes of the economically successful, and against the genes of the poor and the criminal," Clark concludes. "Their extra reproductive success had a permanent impact on the genetic composition of the later population."[9]

Clark's data provide substantial evidence that the English population responded genetically to the harsh stresses of a Malthusian regime and that the shifts in its social behavior from 1200 to 1800 were shaped by natural selection. The burden of proof is surely shifted to those who might wish to assert that the English population was miraculously exempt from the very forces of natural selection whose existence it had suggested to Darwin.

## Evolutionary Changes in China

In China, no equivalent data exist to track changes in social behavior through the generations. But the population clearly fell under intense Malthusian pressure as population density increased. Between 1350 and 1850, the population expanded from 65 million to 430 million. The only checks on growth were the Malthusian constraints of high infant mortality and of malnutrition, which lowered fertility. Female infanticide was a principal means of birth control, with the result that many men could never find wives.

The harshness of the struggle was made no easier by Chinese inheritance practices, which left an estate to be divided equally between the owner's sons. A slightly wealthy peasant family might revert to poverty because each son had to start with a much smaller plot of land. "Each generation, a few who were lucky or able might rise, but a vast multitude always fell, and those families near the bottom simply disappeared from the world," writes the essayist Ron Unz.[10]

A successful family could maintain its economic position over time, Unz writes, "only if in each generation large amounts of additional wealth were extracted from their land and their neighbors through high intelligence, sharp business sense, hard work and great diligence."

Though many poor families perished, there was also movement in the other direction. Within its authoritarian structure, Chinese society was reasonably meritocratic. The examinations for the mandarinate were in principle open to any adult male. Records available from the Ming (1368–1644) and Qing (1644–1912) dynasties show that more than 30% of those who held the highest mandarin rank came from commoner families.

What effect did these forces have on shaping the genetics and social behavior of the Chinese population? There was evidently high selective pressure for survival skills, given that the poorest individuals in each generation were eliminated. Those who worked hard, had the right social skills and made intelligent choices could make their way from the bottom of society to the top in several generations. With a high official's wealth, they could raise more children, amplifying their successful genes before their descendants sank down in status.

Though the Mandarin class might seem at first too small to have exerted any genetic impact on a large population, the examination system operated over many generations and in a population initially much smaller than that of today. The system, though in rudimentary form, was first instituted by the emperor Wu in 124 BC. Over many generations, it would have disseminated upper-class values throughout society as the more numerous children of the well off descended through the social strata.

The examinees were awarded no marks for originality, however. The exams were based on rote memorization of the Chinese classics and formalized commentaries on the text. "It is obvious that such a system of universal examinations, based on examination questions created by a board of senior bureaucrats, established an extraordinary uniformity of attitude and opinion," writes the sociologist of science Toby Huff.[11] The probable effect of the system was to select for excellent memory, high intelligence and unwavering conformity.

At each cycle, the Chinese population became enriched in survival skills. At the same time, authoritarian regimes ruthlessly repressed dissent, just as they do today. This particular set of pressures has borne down on the population for 2,000 years, or some 80 generations, with the evolutionary outcome that has made the Chinese a distinctive population. High intelligence may be one of the behaviors shaped by China's Malthusian regime—Chinese score above Europeans on IQ tests (though so do Koreans and Japanese). Another may be conformity.

# The Long Arc of Domestication

The bourgeoisification of the English population between 1200 and 1800 is a minuscule slice, one that just happens to be documentable, of a long evolutionary process that began in the mists of the last ice age. That process was the civilizing of our remote ancestors, as roving bands of unruly foragers were transformed into people peaceable enough to settle down together.

The process can be called a domestication because, to judge from the evidence of human fossil remains, it seems to parallel the domestication of animal species by the first farmers. As already noted, human skulls and skeletons from about 40,000 years ago become lighter and less robust, as if their owners were no longer fighting one another all the time and could afford more lightly built bone structures.

This lightening of the bone, a genetically based process, is seen in the fossil remains of species like pigs and cattle as they were domesticated from their wild forebears. In people this process, called gracilization, proceeded independently in each of the world's populations, according to the physical anthropologist Marta Mirazón Lahr.[12] All populations followed this trend save for two at the extremities of the human diaspora, the Fuegians at the tip of South America and the aborigines of Australia. Gracilization of the skull is most pronounced in sub-Saharan Africans and East Asians, with Europeans retaining considerable robustness.[13]

In domestic animals, gracilization is one of the side effects of the taming process. The general process is known as pedomorphic evolution, meaning a trend toward the juvenile form. Thus a dog's skull and teeth are smaller than those of a wolf, and the shape of the skull resembles that of a juvenile wolf.

The gracilization of human skulls, the primatologist Richard Wrangham has noted, looks just like the gracilization seen in domestic animals. If this is a side effect of domestication in people too, then exactly who was doing the taming? The obvious answer, Wrangham suggests, is that people must have been taming themselves, by killing or ostracizing individuals who were immoderately violent. Moreover, this ancient process, in his view, is still in motion: "I think that current evidence is that we're in the middle of an evolutionary event in which tooth size is falling, jaw size is falling, and it's quite reasonable to imagine that we're continuing to tame ourselves," Wrangham says.[14] A likely signal of the fact that people today are so much tamer than their forebears is that their shrinking jaws don't now have room for all the teeth that are programmed into them, so the wisdom teeth must often be removed.

Another insight into the human taming process, from a quite different perspective, has been developed by the sociologist Norbert Elias. Despite working in the shadow of the impending Second World War, Elias was fascinated by the decline of violence in Europe since the Middle Ages. He was concerned not with wars between states but with violence in everyday life. He attributed the decline in personal violence to a long-term psychological change in the population, that of the growth of self-restraint.

A starting point for Elias's analysis was medieval treatises on polite manners such as the book *On Civility in Children* by the Renaissance scholar Erasmus. In the 16th century, Europeans' everyday social behavior was beyond gross. It was a social world in which books on good etiquette had to advise people not to blow their noses on the tablecloth nor to snort or smack their lips while eating. People ate with their hands, the fork being a strange luxury. They blew their noses without the aid of a handkerchief or tissues. They performed many bodily functions in public. Their sensibility toward the pain of others was minimal. Public executions were common, often preceded by torture or dismemberment. People behaved with unthinking cruelty toward animals.

A famous midsummer day festivity in 16th century Paris was to burn alive a dozen cats. The king and queen were usually present, and the king or the dauphin would light a pyre. The cats were then tumbled into the flames from an overhead basket, and the crowd reveled in their cries.

"Certainly this is not really a worse spectacle than the burning of heretics, or the torturings and public executions of every kind," Elias writes. "It only appears worse because the joy in torturing living creatures shows itself so nakedly and purposelessly, without any excuse before reason. The revulsion aroused in us by the mere report of the institution, a reaction which must be taken as 'normal' for the present-day standard of affect control, demonstrates once again the long term change of personality structure."[15]

Elias argued that between medieval and modern times, a society-wide shift has taken place toward greater sensibility and more delicate manners. Underlying this civilizing process, he believed, was a psychological shift toward greater self-awareness and self-control. He attributed this change in personality structure in part to the monopolization of force by the state, meaning that individuals needed less to resort to violence in self-defense, and in part to the greater interconnectedness of urban societies, which required individuals increasingly to attune their conduct to that of others and hence to moderate their behavior.

Elias was unable to put numbers on his argument, but these are supplied in profusion in a voluminous survey on violence over the ages by the psychologist Steven Pinker. Contrary to widespread belief that the 20th century was more violent than any other, Pinker establishes that both personal violence and deaths in warfare have been in steady decline for as long as records can tell.

In terms of violence between states, the percentage of people who died in warfare is far higher in pre-state societies, to judge by evidence from archaeology and anthropology, than in the states that succeeded them. The death rate in pre-state societies averages 15% but had fallen to a mere 3% in the first half of the 20th century, a period that includes the two world wars.[16]

Personal violence too has been in steady decline. Between 1200 and 2000, homicide rates per 100,000 people fell from 90 or so to just over one in five European countries.[17] Parallel to the fall in violence is evidence for a general increase in empathy toward the pain of others. People stopped burning women for suspicion of witchcraft; in England the last witch was burned in 1716. Judicial torture was gradually abolished in Europe from 1625 onward.[18] Finally, empathy compelled the abolition of slavery.

Pinker agrees with Elias that the principal drivers of the civilizing process were the increasing monopoly of force by the state, which reduced the need for interpersonal violence, and the greater levels of interaction with others that were brought about by urbanization and commerce.

The next question of interest is whether the long behavioral shift toward more restrained behavior had a genetic basis. The gracilization of human skulls prior to 15,000 years ago almost certainly did, and Clark makes a strong case that the molding of the English population from rough peasants into industrious citizenry between 1200 and 1800 AD was a continuation of this evolutionary process. On the basis of Pinker's vast compilation of evidence, natural selection seems to have acted incessantly to soften the human temperament, from the earliest times until the most recent date for which there is meaningful data.

This is the conclusion that Pinker signals strongly to his readers. He notes that mice can be bred to be more aggressive in just five generations, evidence that the reverse process could occur just as speedily. He describes the human genes, such as the violence-promoting MAO-A mutation mentioned in chapter 3, that could easily be modulated so as to reduce aggressiveness. He mentions that violence is quite heritable, on the evidence from studies of twins, and so must have a genetic basis. He states that "nothing rules out the possibility that human populations have undergone some degree of biological evolution in recent millennia, or even centuries, long after races, ethnic groups, and nations diverged."[19]

But at the last moment, Pinker veers away from the conclusion, which he has so strongly pointed to, that human populations have become less violent in the past few thousand years because of the continuation of the long evolutionary trend toward less violence. He mentions that evolutionary psychologists, of whom he is one, have always held that the human mind is adapted to the conditions of 10,000 years ago and hasn't changed since.

But since many other traits have evolved more recently than that, why should human behavior be any exception? Well, says Pinker, it would be terribly inconvenient politically if this were so. "It could have the incendiary implication that aboriginal and immigrant populations are less biologically adapted to the demands of modern life than populations that have lived in literate state societies for millennia."[20]

Whether or not a thesis might be politically incendiary should have no bearing on the estimate of its scientific validity. That Pinker would raise this issue in a last minute diversion of a sustained scientific argument is an explicit acknowledgment to the reader of the political dangers that researchers, even ones of his stature and independence, would face in pursuing the truth too far.

Turning on a dime, Pinker then contends that there is no evidence that the decline in violence over the past 10,000 years is an evolutionary change. To reach this official conclusion, he is obliged to challenge Clark's evidence that there was indeed such a change. But he does so with an array of arguments that seem less than decisive. Clark's proposed mechanism for the spread of middle-class values is based on the fact that the rich, until recently, had more surviving children than did the poor. Pinker objects that this was true of every society, not just the one that later blasted off into the Industrial Revolution. But this is exactly what Clark's thesis requires to happen in order for the Industrial Revolution to spread to other countries. The mechanism was a pre-condition for the Industrial Revolution wherever it occurred. The specific trigger in England, which explains why it started there rather than in any of the other possible birthplaces in Europe and East Asia, was a sudden boom in the English population.

Pinker notes that countries without a recent history of selection for middle-class values, like China and Japan, can attain spectacular rates of economic growth. But both these countries had long been agrarian economies operating, like England, under Malthusian constraints that favored survival of those who worked hard and saved hard. It was only institutional barriers that delayed these countries' transition to modern economies, and as soon as the barriers were removed, both economies soared. Last, Pinker cites Clark's failure to prove that the English are innately less violent than the inhabitants of countries that have not enjoyed an industrial revolution. This seems an unfair criticism, given that the genes underlying violence are for the most part unknown. Nonetheless, the homicide rate in the United States, Europe, China and Japan is less than 2 per 100,000 people, whereas in most African countries south of the Sahara, it exceeds 10 per 100,000, a difference that does not prove but surely allows room for a genetic contribution to greater violence in the less developed world.[21]

The ultimate proof of Clark's thesis would be discovery of the new alleles that have mediated the social behavior required for Europeans and East Asians to make the transition to modern economies. But there are probably many such genes, each with a small and barely detectible effect, so it may take decades before any come to light.

Meanwhile, his thesis of an evolutionary change provides a powerful explanatory scheme for understanding modern societies, especially when combined with the understanding of political institutions developed by Fukuyama. The countries that have not completed the transition to modern states retain the default state of human political systems, namely that of tribalism.


# Tribal Societies

Africa and much of the Middle East remain largely tribal societies. Tribalism has a bleak reputation because tribal organization is incompatible with that of a modern state. That aside, it is an amazingly ingenious way of securing a rough degree of social order without a government, courts, police force or law books.

In the Arab Middle East, tribalism rests on the idea of group protection. When government does not provide the legal system in which a citizen may seek and secure justice, people rely instead on their relatives. Someone who is wronged will seek his kin's backing against the individual who wronged him. The kin group may be just his family, his extended family or his whole tribe, depending on how far the dispute escalates. A similarly sized group forms around the man he accuses. Hostilities may break out, but the two opposing groups have many inducements to settle, being of generally equal numbers and with each having many members with relatives in the other group. The individual's rights are thus protected, but by the threat of force, not by appeal to law or any formal judicial process.

What makes the system work is the duty felt by each individual to support his group against others, no matter what the personal cost. Failure to support one's family or tribe in a standoff leads to dishonor and means the group's future support of one's own cause may be forfeited.

The tribal system is egalitarian, individualistic and secures redress of wrongs with a minimum of bureaucracy. Despite these outstanding merits, it has grave flaws. It depends on force and group loyalty, not on law. Children are taught from the earliest age that their group is always right and must be supported no matter what. Adults follow the ancient rule: support the nearest group of relatives against the more distant groups. In terms of national politics, the spirit of tribalism leads to "monopoly of power, ruthless oppression of opponents, and accumulation of benefits," writes Philip Salzman, an anthropologist at McGill University who studies nomadic tribes. "In short, it is a recipe for despotism, for tyranny."[22]

The Middle East is, of course, not entirely tribal. It has large urbanized populations. But Middle Eastern governments, in Salzman's view, follow the traditional Ottoman pattern of being largely predatory. They extract taxes from their citizens but provide few services except to protect the mulcted citizenry from other predators. Many people still rely on the tribal system for justice because the government provides none.

The failure to develop modern institutions has led to economic stagnation. There was no economic growth in the region in the 25 years after 1980.[23] Arab countries may have developed the form of Western institutions, but the practice still escapes them. "All Arab countries need to widen and deepen democratic processes to enable citizens to participate in framing public policy on an equal footing," write the Arab authors of a United Nations report on Arab development. "A political system controlled by elites, however decked out with democratic trappings, will not produce outcomes conducive to human security for all citizens," they predict.[24]

The question of why tribalism has persisted in the Middle East but not in Europe has much to do with the nature of the Byzantine, Arab and Ottoman empires that ruled the region for the past two millennia. None was overly concerned with the welfare of its citizens. The Byzantine empire extracted heavy taxes and was widely unpopular, one reason for the success of the Arab empire when the Byzantine empire faltered. The overriding interest of the Umayyad and Abbasid dynasties that followed the Byzantines in ruling the Near East was in steady expansion of Islamic holdings. The Ottoman empire, which eventually displaced the Arabs, was a pure plunder machine. It had continuously to make new conquests and depredations to pay the soldiers on which its imperium depended. Under these circumstances, security of persons and property was consistently low for many centuries. There was nothing equivalent to the steady ratchet mechanism that in England enabled the less violent and more literate to prosper and leave more children like themselves. Because it has always been more rational for its inhabitants to trust the tribe more than the state, tribalism in the Middle East has never disappeared.

In Africa too, tribalism has persisted and interacted poorly with the modern world. Throughout much of Africa, the standard mode of government is kleptocracy; whoever gains power uses it to enrich his family and tribe, which is the way that power has always been used in tribal systems. Extractive institutions, as defined by Acemoğlu and Robinson, are prevalent in Africa, particularly in countries rich in natural resources.

Despite substantial amounts of Western aid, many African countries are little better off than they were under colonial rule. Corruption is rampant. Many services for the poor are siphoned off by elites who leave only a trickle for the intended recipients. Some African countries have lower per capita incomes now than they had in 1980 or, in some cases, in 1960. "Half of Africa's 800 million people live on less than $1 a day," writes the journalist and historian Martin Meredith. "It is the only region where school enrolment is falling and where illiteracy is still commonplace. . . . It is also the only region where life expectancy is falling."

The root of the problem, Meredith believes, is that African leaders have failed to provide effective government. "Africa has suffered grievously at the hands of its Big Men and its ruling elites," he writes.

"Their preoccupation, above all, has been to hold power for the purpose of self-enrichment. . . . Much of the wealth they have acquired has been squandered on luxury living or stashed away in foreign bank accounts or foreign investments. The World Bank has estimated that 40% of Africa's private wealth is held offshore. Their scramble for wealth has spawned a culture of corruption permeating every level of society."[25]

As serious as the flight of capital is the flight of able and educated people. "The most common request to a white visitor to Africa these days, particularly from young people, is for help with a visa to Europe or America. Some 70,000 skilled people are reported to be fleeing the continent each year," writes journalist Patrick Dowden.[26]

Africa south of the Sahara is wracked by frequent violence, with about a third of its countries being at present involved in conflicts. Sudan, since its independence in 1956, has been locked in a series of civil wars. The Congo is a region of unending misery. Nigeria, cursed with oil, is a sea of corruption riven by religious and regional disputes.

Despite all these serious problems, the GNP of the region has recently started to grow, expanding at an average rate of 4.7% a year between 2000 and 2011. Though the increase in GNP masks the extreme inequality that still persists, there are several long-term trends that point to sustainable growth. Overpopulation is not usually regarded as a blessing, yet demographic pressure has played a prime role in the urbanization of Europe and East Asia, though so far not Africa. This, however, may be about to change. "No country or region," say two World Bank economists, Shantayanan Devarajan and Wolfgang Fengler, "has ever reached what the World Bank regards as high-income status with low levels of urbanization. African populations have traditionally been mostly rural, but the cities of sub-Saharan Africa are growing at astonishing rates." Their projection is that in another 20 years, most of the region's population will be urban, as is the case in the rest of the world.[27]

Urbanization and empire building generated the first civilizations in Egypt, Mesopotamia, China and the Americas. Whether it is necessary for Africa to take the same path to create modern states is far from clear. But fierce pressures are clearly at work in the continent, and people will adapt to them. These adaptations may include a reduction of tribalism.

If running a productive, Western-style economy were simply a matter of culture, it should be possible for African and Middle Eastern countries to import Western institutions and business methods, just as East Asian countries have done. But this is evidently not a straightforward task. Though it was justifiable at first to blame the evils of colonialism, two generations or more have now passed since most foreign powers withdrew from Africa and the Middle East, and the strength of this explanation has to some extent faded.

Tribal behavior is more deeply ingrained than are mere cultural prescriptions. Its longevity and stability point strongly to a genetic basis. This is hardly surprising, given that tribes are the default human social institution. The inbuilt nature of tribalism explains why it took so many thousands of years for East Asians and later Europeans to break free of its deadening embrace. It's this escape that is so unexpected, not that the populations of Africa and the Middle East have so far lacked the opportunity to lose the ancient heritage of tribal political behavior.

# The Escape from Tribalism and Poverty

The entry to the modern industrial world has two principal requirements. The first is to develop institutions that enable a society to break away, at least to some substantial extent, from the default human institution of tribalism. Tribalism, being built around kinship ties, is incompatible with the institutions of a modern state. The break from tribalism probably requires a population to evolve such behaviors as higher levels of trust toward those outside the family or tribe. A second required evolutionary change is the transformation of a population's social traits from the violent, short-term, impulsive behavior typical of many hunter-gatherer and tribal societies into the more disciplined, future-oriented behavior seen in East Asian societies and documented by Clark for English workers at the dawn of the Industrial Revolution.

Looking at the three principal races, one can see that each has followed a different evolutionary path as it adapted to its local circumstances. From an evolutionary perspective, no path is better than any other—nature's only criterion for success is how well each is adapted to its local environment.

Consider first Caucasians, the grouping of populations that includes Europeans, Middle Easterners and people of the Indian subcontinent (Indians and Pakistanis). Most European countries followed England almost immediately in transitioning to modern economies. Their populations, like that of England, had abandoned tribalism in the early Middle Ages. Europeans had long lived in the same Malthusian economies that Clark has documented for England. Within a few decades, all had been able to import English production methods and develop modern economies. Thus the Industrial Revolution was not particularly English, given that the evolutionary change that preceded it had occurred throughout Europe and East Asia. For an unrelated reason—the population spurt described above—the Industrial Revolution just happened to manifest itself first in the English economy.

Why did the Industrial Revolution not spread so fast to China or Japan, which differed little from England in the state of their labor, land and capital markets? Clark argues that their upper classes were less fertile than their English counterparts, so that the engine that drove the spread of bourgeois values through the population operated somewhat more slowly in East Asia.[28] The economic historian Kenneth Pomeranz, on the other hand, argues that there were few significant differences between Europe and China until England, with access to the extensive resources in its Caribbean and American colonies, was able to escape the constraints that held China back. He concludes that "forces outside the market and conjunctures beyond Europe deserve a central place in explaining why western Europe's largely unexceptional core achieved unique breakthroughs and wound up as the privileged center of the nineteenth century's new world economy, able to provide a soaring population with an unprecedented standard of living."[29]

From an evolutionary standpoint, the populations of both Europe and East Asia had been primed by the selective pressures of their agrarian economies to escape the Malthusian trap, and it makes little difference which particular factor or event was the trigger for the transition to begin. It seems more likely that institutions, rather than human nature, were the impediment to East Asian progress. The peoples of China, Japan and Korea were fully ready to embrace the Industrial Revolution and market economies once the necessary institutions were in place. For Japan, that was after the Meiji Restoration of 1868; for China, after the reforms initiated by Deng Xiaoping after 1979.

In East Asian populations, history has performed an instructive control experiment with the case of Korea. North and South Koreans are probably very similar to one another genetically, yet North Koreans are poor while South Korea has developed a tiger economy that is post-Malthusian, modern and prosperous. The difference, evidently, lies not in the two countries' genes or geography but in the fact that the same set of social behaviors can support either good or bad institutions. Before 1945, Korea was a single country. After partition, North Korea instituted a collectivist system and a command economy run by a hereditary elite. North Korea lacks property rights or a reliable court system, giving people little incentive to invest for the future, since the state can confiscate property at will. Its population is denied education except for state propaganda. South Korea, by contrast, was directed toward a market economy by its first two authoritarian leaders.

By 2011, South Koreans had become almost 18 times richer than their former compatriots in North Korea, with an estimated GDP per capita of $32,100 compared with $1,800. "Neither culture nor geography nor ignorance can explain the divergent paths of North and South Korea. We have to look at institutions for an answer," say Acemoğlu and Robinson in *Why Nations Fail*.[30]

The fact that China, Japan and South Korea developed modern economies so easily, once the appropriate institutions were in place, is evidence that their populations, like those of Europe, had undergone equivalent behavioral changes to those documented in England.

Another major impact on the Chinese population would have been urbanization. Cities are an environment that rewards literacy, the manipulation of symbols, and high-trust trade networks. With prolonged urbanization, those who mastered the skills of urban living would have had more children, and the population would have undergone the genetic changes that accomplish the adaptation to urban life. In Western countries, the affluent now tend to have fewer children, and China has had its one child policy, both of which set up different evolutionary forces. But until modern times, populations in both Europe and East Asia have partially been shaped by the ability of the rich to raise more surviving children, another instance of the ratchet of wealth.

Turning to the third of the major races, the population of Africa south of the Sahara, the transition of these countries to modern economies has proved considerably slower. Africa is heavily beset with poverty, disease, war and corruption. Despite copious amounts of foreign aid, its living standards have failed to show substantial improvement over those attained under colonial rule. A recent spurt in economic activity in several countries has still not closed the widening gap with East Asia and Europe.

Yet 50 years ago, Africans were as poor as many East Asians. Why has it been easy enough for East Asians to make the transition to modern economies but so hard for Africans?

As discussed, one reason is tribalism. African countries have not developed the institutions to replace tribalism, an essential development for a modern state. African populations have not gone through the same Malthusian wringer that shaped the behavior of the European and East Asian populations. In Africa, population pressure has long been much lower than in Europe and Asia, probably because of poor soils and adverse climates that have restrained food production. State formation, as mentioned, depends on warfare between sizable polities that are forced to compete because of geographical constraints, such as living along a fertile river valley. But intense, large-scale warfare is unlikely to occur until population densities have become so high that people have few other choices.

Until modern times, populations in Africa remained very low, constricted by disease and the grudging fertility of the soil in many tropical regions. For lack of demographic pressure, they thus escaped the urbanization and regimentation to which the populations of Europe and East Asia were subjected for many generations.

From an evolutionary perspective, African populations were just as well adapted to their environment as were those of Europe and Asia to theirs. Small, loosely organized populations were the appropriate response to the difficult conditions of the African continent. But they were not necessarily well suited to the high efficiency economies to which European and East Asian populations had become adapted. From this perspective, it is understandable that African countries should take longer to make the transition to modern economies.

Turning to the Near East, these populations belong, along with Europeans, to the Caucasian grouping. But unlike Europeans and East Asians, they have lacked the shaping experience of living under relatively stable agrarian economies. The Byzantine, Arab and Ottoman empires that held sway in the region for the past 1,500 years were predatory regimes whose purpose was not to serve their populations but to extract wealth from them for the support of the ruling elite. Generations of such rule habituate people, quite rationally, to look to their family and tribe for help, not the government. And under these conditions, it is hard for tribal behaviors to give way to the more trusting behaviors found in modern economies. Countries of the Near East, particularly Arab states, have not yet developed institutions to transcend tribalism and hence face serious obstacles in achieving the transition to modern economies.

These obstacles presumably reside at the level of societies and less with the abilities of their individual members. A worldwide diaspora of accomplished and wealthy Lebanese, for instance, is proof of how successful Lebanese can be outside of Lebanon, a situation similar to that of the overseas Chinese communities during the past two centuries. But China has since been able to improve its institutions more easily than has Lebanon.

# The Problem of Economic Development

The view of economic development generally taken by economists is that people have little or nothing to do with it. Since all humans are identical units that respond the same way to incentives, at least in economic theory, then if one country is poor and another rich, the difference cannot have anything to do with the people but must lie in institutions or access to resources. Just supply enough capital and impose business-friendly institutions, and robust economic growth will surely follow. Strong evidence to this effect seemed to be furnished by the Marshall Plan, which helped revive European economies after the Second World War.

On the basis of this theory, the West has spent some $2.3 trillion in aid over the past 50 years without managing to improve African living standards. Could something be not quite right with the theory? Might the human units of the world's economies be less completely fungible than economic theory assumes, with the consequence that variations in their nature, such as their time preference, work ethic and propensity to violence, have some bearing on the economic decisions they make?

To account for the discrepancy between theory and practice, a few scholars interested in development have begun to suggest that maybe people do matter after all. Their suggestion is that culture plays an important role in people's economic behavior.

In the early 1960s Ghana and South Korea had similar economies and levels of gross national product per capita. Some thirty years later, South Korea had become the 14th largest economy in the world, exporting sophisticated manufactures. Ghana had stagnated, and GNP per capita had fallen to one fifteenth that of South Korea. "It seemed to me that culture had to be a large part of the explanation," the political scientist Samuel Huntington remarked in pondering this divergence of economic fates. "South Koreans valued thrift, investment, hard work, education, organization, and discipline. Ghanaians had different values."[31]

Even the economist Jeffrey Sachs, a tireless advocate of increased aid, has conceded the possibility that culture might play some minor role in differences in economic development. Although "the great divisions between rich and poor countries involve geography and politics," he writes, "nonetheless, there are indeed some hints of culturally mediated phenomena. Two are most apparent: the underperformance of Islamic countries in North Africa and the Middle East and the strong performance of tropical countries in East Asia that have an important overseas Chinese community."[32]

But if culture explains economic performance in even a few groups, it could play a significant role in all economies. Scholars fear pursuing the issue further because they are not really using culture in just its accepted meaning of learned behavior. Rather, it is a catchall word that includes possible reference to a concept they dare not discuss, the possibility that human behavior has a genetic basis that varies from one race to another.

The sociologist Nathan Glazer, for instance, all but admits that culture and race are valid explanatory variables, yet ones that cannot be used: "Culture is one of the less-favored explanatory categories in current thinking. The least favored, of course, is race. . . . We prefer not to refer to or make use of it today, yet there does seem to be a link between race and culture, perhaps only accidental. The great races on the whole are marked by different cultures, and this connection between culture and race is one reason for our discomfort with cultural explanations," he writes.[33]

Several social behaviors that economists have identified as obstacles to progress are ones that could well have a genetic basis. One is the radius of trust, which can extend to strangers in modern economies but is confined to family or tribe in premodern ones. "Seen from the inside, African societies are like a football team in which, as a result of personal rivalries and a lack of team spirit, one player will not pass the ball to another out of fear that the latter might score a goal. How can we hope for victory? In our republics, people outside of the ethnic 'cement' . . . have so little identification with one another that the mere existence of the state is a miracle," writes Daniel Etounga-Manguelle, a Cameroonian economist.[34]

The willingness to save and delay gratification is a social behavior that Clark finds gradually increased in the English population in the 600 years before the Industrial Revolution. Conversely, the propensity to save seems considerably weaker in tribal societies. This could be in large measure because such societies are poorer; everyone saves more as they get richer. But the disinclination to save in tribal societies is linked to a strong propensity for immediate consumption. To quote Etounga-Manguelle again, "Because of the rapport that the African maintains with time, saving for the future has a lower priority than immediate consumption. Lest there be any temptation to accumulate wealth, those who receive a regular salary have to finance the education of brothers, cousins, nephews, and nieces, lodge newcomers, and finance the multitude of ceremonies that fill social life."

There is reasonable evidence that trust has a genetic basis, though whether it varies significantly among ethnic groups and races has yet to be proved. The aspects of culture that some economists have begun to see as relevant to economic performance could well have a genetic basis, even though this has yet to be proved or even seriously investigated. Social behavior, whatever its degree of cultural or genetic foundation, can be modulated by education and incentives, so a better understanding of its role in economic performance might have practical consequences. Those who ignore culture also ignore "an important part of the explanation of why some societies or ethno-religious groups do better than others with respect to democratic governance, social justice, and prosperity," writes the development expert Lawrence Harrison.[35]

The link between race and culture is evident in the well-known natural experiment put in motion by human migrations. Members of various races have migrated to a range of different environments but maintained their distinctive behaviors in many countries over many generations. The economist Thomas Sowell has documented many of these episodes in his trilogy about race and culture.

Consider the case of Japanese immigrants to the United States.

They arrived as agricultural laborers in Hawaii in the late 19th century to work on the sugar crop and later moved to the mainland. The first generation were farmworkers and domestic servants and gained a reputation for hard work. The second generation, with the advantage of American college educations, sought to learn professions. By 1959 Japanese Americans were earning the same family income as European Americans, and by 1990 their income was 45% higher.[36]

In Peru, Japanese workers achieved a reputation for hard work, reliability and honesty and became successful in both farming and manufacturing. In Brazil, Japanese settlers were found to be efficient, industrious and law-abiding. As they prospered, they entered banking and manufacturing and came to own almost 75% as much land in Brazil as there is in Japan. In these three different cultures, the Japanese were successful because of diligent work habits, with the first generation being prodigious farmers and the second generation moving into the professional world.

The overseas Chinese were equally productive immigrants, especially in Southeast Asia, where they worked indefatigably and built up businesses. Most Chinese immigrants began as farm laborers with a prodigious capacity for hard work. In Malaysia, Chinese doing unskilled labor alongside Malays on the rubber plantations would produce twice the output. As early as 1794, a British report on the Malaysian settlement of Penang labeled the Chinese as "the most valuable part of our inhabitants."[37]

Chinese enterprises were typically family owned and family run, even when they grew into sizable corporations. They clung to their own values and work ethic among populations who often took a more relaxed view of how time should be spent. In the Caribbean, Sowell writes, the Chinese "remained outside the value system of West Indian society—unaffected by its Creole patterns of conspicuous consumption, distribution of largesse, forgiveness of debts, and other traits that operate against business success."[38]

Small Chinese populations in Thailand, Vietnam, Laos and Cambodia grew to play disproportionate roles in these countries' economies. They dominated the thriving economy of Singapore and were so productive in Indonesia that their success provoked envy and repeated massacres. By 1994 the 36 million Chinese working overseas produced as much wealth as the 1 billion people in China.[39]

Significant Chinese immigration into the United States began in 1850 with the California gold rush. Often allowed to mine only those areas that others deemed unprofitable, the Chinese persisted and flourished where others couldn't. Chinese workers built much of the Central Pacific Railroad and at one time supplied 80% of all agricultural workers in California.

Their success provoked a series of discriminatory laws advocated by those who could not compete against them. Excluded from one industry after another, by 1920 more than half of all Chinese in the United States worked in laundries and restaurants. As soon as adverse laws were repealed, a younger generation of Chinese Americans started to go to college and enter professional jobs. By 1959, Chinese family income had drawn level with the U.S. average, and by 1990 the median family income was 60% higher than that of non-Asian Americans.[40]

Among non-Asian immigrants, Jews, a special case, are discussed in the next chapter. Germans immigrated to Russia, the United States and Australia, earning a reputation in all three countries for their orderliness and discipline. In Russia they filled many important professions to such a degree that by the 1880s Germans occupied 40% of the Russian army's high command and 57% of the foreign ministry staff. At one time almost the entire membership of the St. Petersburg Academy of Sciences was German.[41]

In the United States, many German immigrants took to farming and were more productive than many other groups. "They were widely known for their industriousness, thrift, neatness, punctuality, and reliability in meeting their financial obligations," Sowell reports. In Australia they became successful farmers, recognized for their hard work, thoroughness and respect for the laws.

The grand theme of Sowell's trilogy is that races have their own strong cultures that shape their behavior, in contrast to the common view that society determines the fates of its minority groups. His purpose is to demonstrate the persistence of racial, ethnic and national cultures but without exploring why such cultural traits endure. He has nothing to say about genetics. But traits that endure, as he has shown, in a range of different environments and from one generation to another are of course quite likely to be anchored by a genetic adaptation; otherwise they would quickly disappear as immigrant groups adapted to their hosts' dominant culture.

Behaviorial traits like industriousness are particularly likely to be retained, but the universal instinct to conform to social rules seems to ensure that the political behaviors of the host country supplant those of the immigrants. Chinese Americans do not organize themselves into authoritarian structures, nor Arab and African Americans into tribal ones.

There is in fact a straightforward explanation for the behaviors of all the migrant groups described by Sowell, in terms of the ratchet of wealth explanation given above for the Industrial Revolution. Populations like Europeans and East Asians, who have adapted, during centuries of living in agrarian systems, to the exigencies of running efficient economies, are at considerable advantage when migrating to other countries. Hard work, efficiency and group cohesion characterize the behavior of East Asian and European migrant groups. It is particularly notable that the Japanese and Chinese should attain higher than average standards of living in the United States, competing against a predominantly European population. The longer history of urbanization in East Asia may underlie part of this competitive advantage.

Populations that have adapted historically to market economies can still fall short of success during periods when they adopt inefficient institutions, such as China under Mao or North Korea under the Kim family dictatorship. When North Korea adopts market-friendly institutions, a safe prediction is that it would in time become as prosperous as South Korea. It would be far less safe to predict that Equatorial Guinea or Haiti needs only better institutions to attain a modern economy; their peoples may not have yet had the opportunity to develop the ingrained behaviors of trust, nonviolence and thrift that a productive economy requires.

## The IQ and Wealth Hypothesis

Standing in sharp contrast to the economists' working assumption that people the world over are interchangeable units is the idea that national disparities in wealth arise from differences in intelligence. The possibility should not be dismissed out of hand: where individuals are concerned, IQ scores do correlate, on average, with economic success, so it is not unreasonable to inquire if the same might be true of countries.

The global IQ/wealth thesis is connected with the interminable debate about black and white IQ differences in the United States, but it involves somewhat different issues and builds more on that part of the evidence about which both sides agree.

The two camps in the IQ debate are known as hereditarians and environmentalists. Both sides generally agree that when IQ tests are administered in the United States, European Americans score 100 (by definition—their scores are normalized to 100), Asian Americans score 105 and African Americans score 85 to 90. The African

American score is noticeably lower than the European score (15 points, or one standard deviation, say the hereditarians; 10 points, say the environmentalists). So much is agreed. The controversy arises in interpreting the gap between the European and African American scores. The hereditarians say the difference in scores is due 50% to environmental reasons and 50% to genetics, although they sometimes change the mix to 20% environment and 80% heredity. The environmentalists assert that the entire gap is due to environmental impediments and that if these were removed, the gap would ultimately disappear altogether.

The heritability of intelligence, the measure which the two sides interpret so differently, does not refer, as easily might be supposed, to the extent to which intelligence is governed by the genes. It refers to the variation in intelligence within a population, and specifically to the extent to which this variation is genetic. A trait could be under complete genetic control, but if there were no variation in the population, its heritability would be zero. Intelligence is almost certainly under genetic influence but none of the responsible alleles has yet been identified with any certainty, probably because each makes too minute a contribution to show up with present methods.[42]

The two sides in the IQ debate are not so terribly far apart on the facts, given that both sides agree that environmental factors are involved. The hereditarians concede that if an adjustment is made for socioeconomic status, with which IQ score is correlated, then African American scores would rise 5 points, to 90. That is not so much greater than the gap that separates Asian Americans from European Americans, about which no one seems to be bothered.

Why, then, is the debate so heated? The acrimony arises because the two positions lead to different policy choices. The hereditarians say that since the IQ gap is substantially innate, the Head Start early education program has failed, as was predicted by Arthur Jensen in 1969, and so will similar interventions. The environmentalists deny this, saying the gap in educational attainment is closing, and that it is the racist nature of society that impedes African American advancement.

That issue needn't be resolved here. The question of global IQ is a somewhat less fraught issue and is of considerable evolutionary interest because intelligence reflects evolutionary changes in the brain and behavior.

The principal proponents of the global IQ/wealth thesis are Richard Lynn, a psychologist at the University of Ulster, and Tatu Vanhanen, a political scientist at the University of Tampere in Finland. They have gathered data from around the world and worked out the correlation between intelligence, as measured by IQ tests, and various criteria of economic success, such as gross national product per capita. Their findings are published in two books, *IQ and the Wealth of Nations* (2002) and *IQ and Global Inequality* (2006).

The world's average IQ, they report, is 90. Broken down by race, the IQ of East Asian nations is 105, the European score is 99, and sub-Saharan Africa's is 67.[43] The authors note that the sub-Saharan African score would be considerably higher but for malnutrition and ill health.

Lynn and Vanhanen argue that IQ scores must be measuring something significant because IQ correlates well with measures of educational attainment. The scores are indeed strongly associated, they say, with what economists call human capital, which includes training and education.

Turning to economic indicators, they find that national IQ scores have an extremely high correlation (83%) with economic growth per capita and also associate strongly with the rate of economic growth between 1950 and 1990 (64% correlation).[44]

"Our argument is that differences in the average mental abilities of populations measured by national IQ provides the most powerful, although not complete, theoretical and empirical explanation for many types of inequalities in human conditions," Lynn and Vanhanen conclude.[45] It follows from this conclusion that not much can be done to reduce inequities in national wealth. "The gap between rich and poor countries can be expected to persist as far as it corresponds to differences in national IQs," they say.[46]

It may seem intuitively plausible that a more intelligent population might garner more wealth than a less intelligent one. But intelligence is a quality of individuals, not of societies. A society of strong men might easily be defeated by weaker men if the weaklings are more cohesive and fight harder. Like strength, the property of individual intelligence does not necessarily transfer from individuals to the society of which they are part.

And indeed with Lynn and Vanhanen's correlations, it is hard to know which way the arrow of causality may be pointing, whether higher IQ makes a nation wealthier or whether a wealthier nation enables its citizens to do better on IQ tests. The writer Roy Unz has pointed out from Lynn and Vanhanen's own data examples in which IQ scores increase 10 or more points in a generation when a population becomes richer, showing clearly that wealth can raise IQ scores significantly. East German children averaged 90 in 1967 but 99 in 1984. In West Germany, which has essentially the same population, averages range from 99 to 107. This 17 point range in the German population, from 90 to 107, was evidently caused by the alleviation of poverty, not genetics.

There is a 10 to 15 point difference in IQ scores between the richer and poorer countries of Europe, yet these differences disappear when the inhabitants migrate to the United States, so the differences are evidently an environmental effect, not a genetic one. If European IQ scores can vary so widely across different decades and locations, it is hard to be sure that any other ethnic differences are innate rather than environmental. Lynn and Vanhanen's book "constituted a game-ending own-goal against their IQ-determinist side," Unz concluded, but "neither of the competing ideological teams ever noticed."

Lynn and Vanhanen do in fact acknowledge the role of wealth

in enhancing IQ scores. But the difficulty of quantifying the IQ-enhancing effect of wealth seriously weakens the ability of IQ scores to explain wealth. More generally, it may be hazardous to compare the IQ scores of different races if allowance is not made for differences in wealth, nutrition and other factors that influence IQ.

East Asia is a vast counterexample to the Lynn/Vanhanen thesis. The populations of China, Japan and Korea have consistently higher IQs than those of Europe and the United States, but their societies, despite their many virtues, are not obviously more successful than those of Europe and its outposts. Intelligence can't hurt, but it doesn't seem a clear arbiter of a population's economic success. What is it then that determines the wealth or poverty of nations?

## Institutions and National Failure

A much praised inquiry into the nature of national poverty is the recent book *Why Nations Fail,* by Daron Acemoğlu, an economist, and James Robinson, a political scientist. As noted earlier in this chapter, they agree with Fukuyama in regarding institutions as critical to understanding how human societies work. And they arrive at this conclusion by an independent route. Fukuyama identifies the role of institutions largely through historical patterns; Acemoğlu and Robinson emphasize political and economic analysis.

Most of the inequality between the countries of the world has emerged since the Industrial Revolution, Acemoğlu and Robinson note, before which time standards of living were almost uniformly low for almost everyone except for the handful of people in each nation's ruling class. A list of the 30 richest countries today would include Britain and the countries to which the Industrial Revolution quickly spread—Western Europe and the initially British settlements of the United States, Canada and Australia—and Japan, Singapore and South Korea. The 30 poorest countries would be mostly in sub-Saharan Africa, joined by Afghanistan, Haiti and Nepal. Going back a century, the list of countries in the top and bottom 30 would be much the same, save that Singapore and South Korea hadn't joined the ranks of the richest.

Surely economists, historians or other social scientists must have devised some convincing explanation for this substantial and enduring inequality? "Not so," say Acemoğlu and Robinson: "Most hypotheses that social scientists have proposed for the origins of poverty and prosperity just don't work and fail to convincingly explain the lay of the land."[47]

Their thesis is that there are bad and good institutions or, as they term them, extractive and inclusive institutions. The bad, extractive institutions are those in which a small elite extorts the most it can from a society's productive resources and keeps almost everything for itself. The elite opposes technological change because it is disruptive of the political and economic order required to maintain their position. Through its own greed, the elite impoverishes everyone else and prevents progress. A permanent vicious circle between the society's extractive political and economic institutions maintains continual stagnation.

Good, inclusive institutions, by contrast, are those in which political and economic power is widely shared. The rule of law and property rights reward endeavor. No sector of society is powerful enough to block economic change. A virtuous circle between politics and economics maintains increasing prosperity.

The archetype of inclusive institutions, in Acemoğlu and Robinson's view, was the Glorious Revolution of 1688, in which England replaced its French-leaning king, James II, with his son-in-law William of Orange, a switch that consolidated Parliament's control over the king. Both political and economic institutions became more inclusive, creating incentives for entrepreneurs and laying the basis for the Industrial Revolution.

This shift to inclusive institutions was so decisive, in Acemoğlu and Robinson's view, that it is in fact the only condition that distinguishes rich countries from poor ones. Comparing England and Ethiopia, one of the world's richest countries and one of the poorest, they assert that "the reason Ethiopia is where it is today is that, unlike in England, in Ethiopia absolutism persisted until the recent past."[48]

They concede that absolutist regimes can generate prosperity for a while, for instance by switching manpower from agriculture to industry. But these one-off expedients were temporary in the case of the Soviet Union; and in China too, political repression will also, they predict, cause the Chinese economy to falter unless it makes its political institutions more inclusive.

If inclusive institutions are the only thing that matters in achieving prosperity, it follows that foreign aid is useless unless it begins with institutional reform. But this is almost never the case, because such conditions are resisted by the ruling elites whose interests would be imperiled by the reforms. As Acemoğlu and Robinson explain, "Countries need inclusive economic and political institutions to break out of the cycle of poverty. Foreign aid can typically do little in this respect, and certainly not with the way that it is currently organized."[49]

As a description of the current state of affairs, Acemoğlu and Robinson's thesis seems reasonably satisfying. But the authors have great difficulty explaining how good institutions arise or how they can be established in a country that doesn't have them. "The honest answer of course is that there is no recipe for building such institutions," they admit.

They have no recipe to offer because they believe that good institutions have emerged as a matter of chance, as random ripples on the inexplicable tides of history. They argue that institutions change because of "institutional drift," a phenomenon they explicitly compare to the random process of genetic drift. They think that institutions are shaped by history but that history moves in a "contingent path," meaning that it is a succession of accidents. Even the Glorious Revolution was not inevitable, since its emergence "was in part a consequence of the contingent path of history."[50]

Acemoğlu and Robinson argue that bad institutions get replaced with good ones, as in England's Glorious Revolution or Japan's Meiji Restoration, because of "critical junctures" in history combined with "propitious existing institutions." They assert, "In addition some luck is key, because history always unfolds in a contingent way."[51]

Luck is an explanation? Not divine providence, or some sign of the zodiac? The authors are driven to reach for such unsatisfying explanations because they have ruled out the obvious possibility that variations in human behavior are the cause of good or bad institutions. They are thus forced back on nonexplanations like luck and the contingent path of history.

The wealth of human societies has not followed some random path over the past millennium, but rather, as Acemoğlu and Robinson observe, a part of the world has grown steadily and vastly richer over the past 300 years. This is not an accident or luck, and a reasonable explanation is available in terms of human evolution.


## The Meso-Industrial Age

The explanation is that there has been an evolutionary change in human social behavior that has facilitated the new, post-tribal social structure on which modern societies are based. Rich countries have non-tribal, trust-based economies and favorable institutions. Poor countries are those that have not fully escaped from tribalism and labor under extractive institutions that reflect their limited radius of trust.

The present world situation is analogous to the mixed social structures that prevailed during the Mesolithic Age, which lasted from about 10,000 to 5,000 years ago in Europe. People using the new farming technologies had begun to invade Europe from the Near East. The hunter-gatherer people who then occupied Europe were either killed or adopted into the new farming communities. The hunter-gatherers used an old kit of stone tools, which archaeologists refer to as Paleolithic, in contrast to the new kit used by the farmers, known as Neolithic. The transition period from the Paleolithic to the Neolithic, during which the settled behavior became increasingly dominant in Europe, is therefore known as the in-between or Mesolithic Age.

The world is at present in a similar transition period, in which some populations have emerged from the shaping forces of Malthusian agriculture and others are still in the throes of the process. The Meso-Industrial Age, as it might be called, is the period during which the rest of the world, principally the countries of sub-Saharan Africa and the Middle East, makes the evolutionary transition to the social behaviors needed to support modern economies. No doubt the process requires some adaptation and a change in institutions. But given the speed of evolution and the rapidity of cultural change in today's world, the Meso-Industrial Age may be over in far fewer generations than might be expected.

It is now time to consider a special population that for many centuries lacked a homeland of its own. Jewish culture is as distinctive as that of many other groups but, because of its particular nature, a strong case can be made that in important respects its culture has genetic roots.

# 8

---

# JEWISH ADAPTATIONS

Surely . . . Judaism is more than the history of anti-semitism. Surely Jews deserve to be defined—and are in fact defined, by others as well as by themselves—by those qualities of faith, lineage, sacred texts and moral teachings that have enabled them to endure through centuries of persecution.

—GERTRUDE HIMMELFARB[1]

In many spheres of life, Jews have made contributions that are far larger than might be expected from their numbers. Jews constitute 0.2% of the world's population, but won 14% of Nobel Prizes in the first half of the 20th century, despite social discrimination and the Holocaust, and 29% in the second. As of 2007, Jews had won an amazing 32% of Nobel Prizes awarded in the 21st century.[2]

Jews have excelled not only in science but also in music (Mendelssohn, Mahler, Schoenberg), in painting (Pissarro, Modigliani, Rothko), and in philosophy (Maimonides, Bergson, Wittgenstein). Jewish authors have won the Nobel Prize in Literature for writing in English, French, German, Russian, Polish, Hungarian, Yiddish and Hebrew.[3]

Such achievement requires an explanation, and the best and simplest is that Jews have adapted genetically to a way of life that requires higher than usual cognitive capacity. People are highly imitative, and if the Jewish advantage were purely cultural, such as hectoring mothers or a special devotion to education, there would be little to prevent others from copying it. Instead, given the new recognition of human evolution in the historical past, it is more likely that Jewish intellectual achievement has emerged from some pressure in their special history. Just as races have evolved in the recent past, ethnicities within races will also evolve if they are reproductively isolated to some extent from their host population, whether by geography or religion. The adaptation of Jews to a special cognitive niche, if indeed this has been an evolutionary process, as is argued below, represents a striking example of natural selection's ability to change a human population in just a few centuries.

Until the era of rapid DNA sequencing, it could be surmised that Jews were a distinct population because of religious laws that frowned on marriage outside Judaism. But no one knew for certain because in the absence of genetic evidence it was impossible to estimate the amount of intermarriage that might nevertheless have occurred throughout history. DNA analysis shows that Jews are a definable set of populations and that Ashkenazi Jews, at least, can be distinguished genetically from other Europeans. With each Jewish community, there has been some intermarriage with local populations but at a very slow rate. This neatly explains the observation by Jewish anthropologists that Jews from all over the world resemble one another yet also resemble their host populations.

The basis of the common resemblance is that Jews originated in Israel and carry shared inheritance from the Semitic population of the region. As recently as 3,000 years ago, a date that marks the probable beginning of the Jewish religion, Jews were no different from anyone else: they were part of the general Near East population from which today's Arabs, Turks and Armenians are also descended. But as soon as their religion started forbidding members to marry nonmembers, the Jewish population would have entered into reproductive isolation, much as if it had been placed on a remote island. Some large degree of reproductive isolation is the necessary condition for a population to take its own evolutionary path.

As to European Jews, or Ashkenazim, genetics show that there has been a 5% to 8% admixture with Europeans since the founding of the Ashkenazi population in about 900 AD, which is equivalent to 0.05% per generation.[4] Researchers using a SNP chip that tests the genome at 550,000 sites report that they were able to distinguish with complete accuracy between Ashkenazim and non-Jewish Europeans. This is a test applicable to populations, not individuals, since it depends on seeing how individuals cluster together in terms of statistical differences in their genome sequences. Still, it shows that Ashkenazim are a distinctive population and therefore could have been subjected to forces of natural selection different from those acting on other Europeans.

Ashkenazim probably differ genetically from other Europeans because of the Near Eastern component in their ancestry. "It is clear that the genomes of individuals with full Ashkenazi Jewish ancestry carry an unambiguous signature of their Jewish heritage, and this seems more likely to be due to their specific Middle Eastern ancestry than to inbreeding," the researchers say.[5]

The rate of admixture with host populations has probably been similar among the other two main Jewish populations. These are the Sephardim and the Oriental Jews or Mizrahim. The Sephardim are Jews who had long lived in Spain and Portugal but were expelled from those countries in 1492 and 1497. They then dispersed around the Mediterranean to places like North Africa and the Ottoman empire. Many Sephardim also settled in Holland. Oriental Jews are those who have long lived in Arab countries and Iran. The origin of the Sephardim is still obscure but there are genetic hints that both they and the Ashkenazim may be offshoots of the large Jewish community that lived in Rome during the early Roman empire.

On genetic maps of the world's population, the three Jewish groups cluster together, sandwiched between the Middle Eastern populations with whom they share joint ancestry and the European populations with which the Ashkenazim and Sephardim are admixed.

Given this degree of genetic separation, it is perfectly possible for Jewish populations to have followed a slightly different evolutionary path from Europeans as they adapted to the special circumstances of their history and developed unusual cognitive abilities.

Yet the idea that there could be meaningful genetic differences between human groups is fiercely resisted by many researchers. They cling to the idea that the mind is a blank slate on which only culture, not genetics, can write, and dismiss the possibility that evolution could have effected any recent change in the human mind. They reject the proposal that any human behavior, let alone intelligence, has a genetic basis. They make accusations of racism against anyone who suggests that cognitive capacities might differ between human population groups. All these positions are shaped by leftist and Marxist political dogma, not by science. Nonetheless, most scholars will not enter this territory from lively fear of being demonized by their fellow academics.

A more substantive objection to exploring this issue has to do with sensitivities of the Jewish community. As with the Chinese immigrant communities in Asia, hard work and success has too often provoked the envy and enmity of their host populations, leading to discrimination, expulsions and massacres. Discussion of Jewish intelligence carries the risk of stirring up hostility. But the days of pogroms are past, and to ignore every difficult subject would serve only the forces of obscurantism.

The only serious recent attempt by researchers to delve into the links between Jewish genetics and intelligence is an extended essay by Gregory Cochran, Jason Hardy and Henry Harpending of the University of Utah. Their report was submitted to several journal editors in the United States, all of whom said it was fascinating but that they could not publish it. The authors eventually secured publication in England, in the *Journal of Biosocial Science*.[6]

The essence of the Utah team's argument is to assert a causal connection between two unusual and otherwise unexplained facts. The first is that Ashkenazi Jews, in addition to their cultural achievements, have a high IQ—generally measured at between 110 and 115, which is the highest average of any ethnic group. The second is that Ashkenazim have a strange pattern of so-called Mendelian diseases, those that are caused by a mutation in a single gene.

The Utah researchers note first that Ashkenazi IQ, besides being high, has an unusual structure. Of the components of IQ tests, Ashkenazim do well on verbal and mathematical questions but score lower than average on visuospatial questions. In most people, these two kinds of ability are highly correlated. This suggests that some specific force has been at work in shaping the nature of Ashkenazi intelligence, as if the population were being adapted not to hunting, which requires excellent visuospatial skills, but to more urban occupations served by the ability to manipulate words and numerals.

So it's striking to find that Ashkenazim, almost from the moment their appearance in Europe was first recorded, around 900 AD, were heavily engaged in moneylending. This was the principal occupation of Jews in England, France and Germany. The trade required a variety of high level skills, including the ability to read and write contracts and to do arithmetic. Literacy was a rare ability in medieval Europe. As late as 1500, only 10% of the population of most European countries was literate, whereas almost all Jews were.[7]

As for arithmetic, it may be simple enough with the Arabic numerals in use today. But Arabic numerals did not become widespread in Europe until the mid-16th century. Before that, people used Roman numerals, a notation system that has no zero. Calculating interest rates and currency swaps without the use of zero is not a straightforward computation.

There were no banks in those days, and moneylenders were essential for those who wished to buy on credit or engage in long-distance trade. The moneylender had to assess the creditworthiness of borrowers, appraise collateral, understand local contract law, and stay on good terms with the authorities who would enforce it. For those engaged in long-distance trade, in which physical transfer of money was generally avoided because of the danger, it was necessary to arrange credit with reliable partners in faraway cities.

So it's easy enough to accept the Utah team's first premise, that Ashkenazi Jews of the Middle Ages were engaged in a cognitively demanding occupation. The second point is that this occupation, though highly risky, was also highly rewarding. In all the European countries in which they settled, Jews enjoyed high standards of living. Between 1239 and 1260, taxes paid by Jews contributed between one sixth and one fifth of royal revenues, even though Jews constituted 0.01% of the population. In 1241 Jews in Germany paid 12% of the entire imperial tax revenue.[8]

The wealth was important because it enabled Jews to secure a considerable degree of reproductive success. Before the Industrial Revolution and the escape from the Malthusian trap, the rich had more surviving children, being able to provide better nutrition and warmer houses. The Ashkenazi population had grown from almost nothing in 900 AD to about 500,000 people by 1500 AD, and had reached 14.3 million by 1939.[9]

From about 900 AD to 1700 AD, Ashkenazim were concentrated in a few professions, notably moneylending and later tax farming (give the prince his money up front, then extract the taxes due from his subjects). Because of the strong heritability of intelligence, the

Utah team calculates that 20 generations, a mere 500 years, would be sufficient for Ashkenazim to have developed an extra 16 points of IQ above that of Europeans. The Utah team assumes that the heritability of intelligence is 0.8, meaning that 80% of the variance, the spread between high and low values in a population, is due to genetics. If the parents of each generation have an IQ of just 1 point above the mean, then average IQ increases by 0.8% per generation. If the average human generation time in the Middle Ages was 25 years, then in 20 human generations, or 500 years, Ashkenazi IQ would increase by $20 \times 0.8 = 16$ IQ points.

There were of course Christian moneylenders who required the same cognitive skills as Ashkenazim. But the Christians married into a much larger community that included people in many other occupations. Natural selection may have been raising the intelligence of urban populations in general during the Middle Ages but exerted a much stronger effect on the smaller Jewish population. This was because any intelligence-enhancing genes that arose in a family in the general population would be diluted in the next generation, but could accumulate in the Jewish community because marriage to outsiders was deterred. This selective effect could not operate on Oriental Jews—those under Muslim rule—because their rulers for the most part confined them to unpopular occupations like tanning or butchery which required no particular intellectual skills. Oriental Jews and Sephardim are not overrepresented in cognitively demanding occupations and both have IQs comparable to Europeans, the Utah team says.

The Utah researchers give short shrift to the other explanations that have been proposed for enhanced Jewish intelligence. One is that the series of massacres and expulsions that began at the time of the First Crusade in 1096 constituted a selective effect that only the more intelligent were able to survive. But the massacres and expulsions affected the whole Ashkenazi population and seem unlikely to have selected the more intelligent nearly as precisely as did the skills required for moneylending.

Jewish folklore holds that marriages between the children of rabbis and rich merchants were the driver of enhanced intelligence. Talmudic academies, writes the anthropologist Melvin Konner, "culled the best minds in every generation of Jews for more than a thousand years. Rising stars among these bright young men would board with successful merchants, and matches would be made between them and the merchants' daughters. Thus the smartest, most studious boys would join the wealthiest families."[10]

Without any data as to how often such matches were made, this seems more like a scholar's fantasy than a common arrangement. Rich merchants may have been more likely to see another merchant's son as a more promising son-in-law than a poor rabbinical student. But even if such marriages did sometimes occur, there were not enough rabbis in the population—a mere 1%—to make a genetically significant difference, the Utah team says.

The Utah researchers make a plausible enough general case that the selective pressure from a cognitively demanding occupational niche would have selected for higher intelligence among Ashkenazim. They then go on to identify what they believe are the causative genes. Their proposals, if confirmed, would give specific plausibility to the general argument but, if false, would not bring it down.

The genetic argument concerns the mutations that cause Mendelian diseases. Mendelian, or simple, diseases are those that result from a mutation that disables a single gene, as opposed to complex diseases like cancer or diabetes, which are the product of several causative variant genes.

Every population has its own pattern of Mendelian diseases. Among Jews, some Mendelian mutations, like familial Mediterranean fever, are very ancient, being shared with other Middle Eastern populations like Turks and Druze, while others are found among only Ashkenazim or Sephardim and so must have occurred after the two populations separated.

The Utah team's analysis focuses on a group of four Mendelian diseases that occur in Ashkenazim and affect an obscure biochemical function, the storage of fats known as sphingolipids. The four diseases are known as Tay-Sachs, Gaucher's, Niemann-Pick and mucolipidosis type IV.

Inheriting a single copy of any of these variant genes does no great harm: the good copy inherited from the other parent compensates for the defective allele. But inheriting a double dose of the variant alleles can cause serious impairment in the case of Gaucher's and is lethal in the case of the other three diseases.

The variant genes that cause the four diseases are found in relatively high proportions in the Ashkenazi population. When a version of gene is more common than expected, geneticists usually assume one of two causes. One is natural selection and the other is the influence known as a founder effect.

Why should natural selection favor a variant gene associated with a lethal disease? This can happen when the variant, though lethal in a double dose, confers some advantage when inherited from only one parent. A well-known example is that of sickle-cell anemia. A person with one copy of the variant gene is protected from malaria, but those who inherit two copies suffer from a serious blood disease. The allele will be favored by natural selection because the many single-allele carriers, who are protected from malaria, far outnumber the carriers of two alleles, who die or suffer impairment.

The other reason why a variant gene can be more common than expected is that it happened to occur at high frequency in a small population that later expanded. Any rare mutation carried by one of the population's founders will be inherited by his or her descendants and attain a higher frequency in that population than in most others, a situation known as a founder effect.

The geneticist Neil Risch has concluded that the Ashkenazi Jewish mutations are founder effects that arose around 1,000 years ago. Since the mutations all arose at the same time, they must have the same cause, and that must be a founder effect, Risch argues, because such a variety of mutations is unlikely to offer any specific advantage that natural selection might favor.[11]

But this argument is neatly turned around by the Utah team. They agree that the Ashkenazi mutations arose within the past 1,000 years but argue that the mutations were indeed all favored by natural selection at the same time because they all promote intelligence.

If the founder effect argument is rejected, a plausible reason for the commonness of Ashkenazi Mendelian mutations would be that they protect against some serious disease. But no such protective effect has yet been detected. In any case, Ashkenazi Jews and the European populations they lived among suffered from the same diseases, yet there is no similar pattern of mutations in Europeans.

The only significant difference in the Ashkenazi way of life was that they worked in cognitively demanding occupations, the Utah team argues, so this must be the selective pressure that drove the Ashkenazi Mendelian mutations to such relatively high levels.

Another reason for assuming natural selection is at work, rather than a founder effect, is that some of the Ashkenazi mutations occur in clusters. This is highly unusual because mutations strike at random throughout the genome so should not be concentrated in genes that all have the same function. One set of Ashkenazi mutations occurs in the cluster of genes that controls the sphingolipid storage pathway mentioned above. For four mutations to be found in a specific pathway is a strong indication of natural selection. The Utah team points to experimental evidence, though there is not very much of it, suggesting that disruption of sphingolipid storage induces neurons to make more connections than usual.

A second cluster of four mutations is found in a DNA repair pathway. Two of the mutations occur in the BRCA1 and BRCA2 genes and are associated with breast and ovarian cancer. The other two mutations cause the diseases Fanconi's anemia type C and Bloom syndrome. It is hard to see how disruptions of DNA repair systems could be beneficial in any context, especially in the case of the two BRCA mutations, which carry risk even when an individual has a single copy of the mutated gene. The Utah team notes that BRCA1 can limit cell proliferation in neuronal stem cells in the embryo and adult, so that impairing the gene could allow extra brain cells to be generated. They suggest there may be similar advantages, yet to be discovered, in the other DNA repair mutations.

Though the exact role of the Mendelian mutations in promoting intelligence has yet to be clarified, they are strikingly common among Ashkenazim. Some 15% of Ashkenazi Jews carry one of the sphingolipid or DNA repair mutations, and 60% carry either these or one of the other Mendelian disease mutations special to Ashkenazim. As already noted, the mutations are harmless when inherited from just one parent. The Utah team's explanation seems the best so far for this strange pattern of mutations, and in particular for those that exist in clusters. Moreover, it is a great virtue in a scientific hypothesis to be easily testable, as the Utah team's theory is. The theory implies that people carrying one of the Ashkenazic mutations will be found to have higher IQ scores, on average, than people who do not. Anyone with access to a population of Ashkenazim could test the prediction that high IQ is associated with the Ashkenazic mutations. Strange to say, no one has yet done so or, if they have, they have not published their findings.

Without being able to test a living population, the Utah team obtained indirect evidence that Gaucher's disease raises IQ. They found that of 255 working age patients at a Gaucher's clinic in Israel, one third were in fields like science, accountancy or medicine, which require high IQs, a far greater proportion than in the population as a whole.

## Advantages of Literacy

A possible weakness of the Utah team's proposal is the assertion that enhanced cognitive capacity is confined to the Ashkenazic branch of the Jewish population. Sephardim have given the world Spinoza, Disraeli, Ricardo and many other distinguished individuals. It is hard to find specific measurements of Sephardic IQ, and the Utah team offer none in their article. Measurements of IQ in Israel report that Ashkenazi IQ is higher than that of non-Ashkenazim, but the latter group includes Oriental Jews as well as Sephardim. The Utah team focuses on Ashkenazi Jews because the Mendelian mutations found in Ashkenazim seem to have originated around 1000 AD, after Ashkenazim and Sephardim became separate populations. But even if the Utah team's thesis has merit, there is no reason why Jews should not have enjoyed special cognitive capacities from much earlier in their history; if so, these traits could later have been enhanced among Ashkenazim in the manner the Utah team describes.

A new perspective on Jewish history has recently been developed by two economic historians, Maristella Botticini and Zvi Eckstein. Botticini specializes in medieval contracts and marriage markets and teaches at Bocconi University in Milan. Eckstein is a distinguished economist who has served as deputy governor of the Bank of Israel. Their interest in Jewish history is focused on population numbers and occupation. They allude hardly at all to intelligence or genetics, yet their economic history makes abundantly clear how selection pressures could have acted on the Jewish population so as to enhance cognitive capacity.

The widely held conventional explanation for Jewish occupational history is that Jews were forbidden by their Christian host nations to own land and drifted into moneylending because it was the only profession open to them. Because of frequent expulsions and persecutions, according to this view, Jewish communities were dispersed in towns all over Europe and the Mediterranean world.

Botticini and Eckstein reject this explanation, arguing with a wealth of historical detail that Jews were not forced into moneylending but rather chose it because it was so profitable, and that they generally dispersed not because of persecution but because there were jobs for only so many moneylenders in each town.

But how did Jews come to choose this unusual occupation? Botticini and Eckstein develop a simple but forceful explanation that goes back to the beginnings of rabbinical Judaism in the 1st century AD.

Before the rabbinical era, Israelite religion was focused on the temple in Jerusalem and on copious animal sacrifices. Its leaders promoted three major insurrections against Roman rule, the first of which culminated in the destruction of the Jerusalem temple in 70 AD. The loss of the temple strengthened the position of the Pharisees, one of several sects, and led them to develop a quite different version of Judaism in which the temple and animal sacrifice were replaced as central components of the religion by study of the Torah.

The rabbinical form of Judaism that emerged from this movement emphasized literacy and the skills to read and interpret the Torah. Even before the destruction of the temple, the Pharisee high priest Joshua ben Gamla issued a requirement in 63 or 65 AD that every Jewish father should send his sons to school at age six or seven. The goal of the Pharisees was universal male literacy so that everyone could understand and obey Jewish laws. Between 200 and 600 AD, this goal was largely attained, as Judaism became transformed into a religion based on study of the Torah (the first five books of the Bible) and the Talmud (a compendium of rabbinic commentaries).

This remarkable educational reform was not accomplished without difficulty. Most Jews at the time earned their living by farming, as did everyone else. It was expensive for farmers to educate their sons and the education had no practical value. Many seem to have been unwilling to do so because the Talmud is full of imprecations against the *ammei ha-aretz,* which in Talmudic usage means boorish country folk who refuse to educate their children. Fathers are advised on no account to let their daughters marry the untutored sons of the *ammei ha-aretz.*

The scorned country folk could escape this hectoring without totally abandoning Judaism. They could switch to a form of Judaism Lite developed by a diaspora Jew, one that did not require literacy or study of the Torah and was growing in popularity throughout this period. The diaspora Jew was Paul of Tarsus, and Christianity, the religion he developed, seamlessly wraps Judaism around the mystery cult creed of an agricultural vegetation god who dies in the fall and is resurrected in the spring.[12]

As evidence that many Jews did indeed convert to Christianity, Botticini and Eckstein cite estimates showing that the Jewish population declined dramatically from around 5.5 million in 65 AD to a mere 1.2 million in 650 AD. There is little else to account for such a dramatic decline other than a high rate of conversion away from Judaism.

Botticini and Eckstein make no mention of the genetic forces that would have been brought into play by such a conversion. But if Jews who lacked the ability or commitment to become literate were shed from the community generation after generation, the propensity for literacy of those remaining would steadily rise. The rabbinical requirement for universal male literacy may thus have been the first step toward a genetic enhancement of Jewish cognitive capacity. A second step was to come later, when the literacy was put to great practical effect.

By 650 AD, Jews had almost entirely disappeared from regions where Christianity was strong, including Syria, Lebanon, Egypt and even Israel itself. The center of the Jewish world shifted to Iraq and Persia. There was also a shift in Jewish occupations. Jews abandoned farming and moved to towns, where they entered trade and commercial activities or became shopkeepers and artisans.

After the establishment of the Abbasid caliphate in 750 AD, Jews migrated to the newly prospering towns and cities. By 900 AD, almost all Jews were engaged in urban occupations, dealing with crafts, trade, moneylending and medicine. Why did Jews choose professions in these particular fields? Common belief is they were forbidden to own land and denied entry to certain crafts. Botticini and Eckstein say there is little or no evidence for such prohibitions. Jews concentrated in professions like trade and moneylending, they argue, for a simple reason. In a world where most people were illiterate, the literacy of almost all Jews gave them a decided advantage in any occupation that required reading contracts or keeping accounts.

Jews enjoyed another practical benefit conferred by their religion. Jewish communities were subject to law, as laid out in the Talmud, and rabbinic courts oversaw contract enforcement and disputes. Because of the presence of Jewish communities in many cities of Europe and the Near East, Jews had access to a natural trading network of their coreligionists. Both the network and the dispute resolution mechanism were unusual and gave Jews a special advantage in long-distance commerce.

As trade and urbanization started to flourish in the Muslim world under the Abbasids, the "higher literacy of the Jewish people," Botticini and Eckstein write, "gave the Jews a comparative advantage over non-Jews in crafts, trade, commerce and money-lending."[13]

The sack of Baghdad by the Mongols in 1258 destroyed the political and cultural center of the Abbasid empire, and large regions of Iraq and Persia became depopulated. The population center of the Jewish community now shifted to Europe, where Jews increasingly specialized in moneylending.

This occupational pattern had a profound demographic consequence. Because moneylending was so profitable, despite its high risks, Jews could afford to support large families and, like other wealthy people, could ensure that more of their children survived to adulthood. After the devastation of the Jewish communities in Iraq and Persia and the expulsion of European Jews from England, France and many regions of Germany, their total population fell to fewer than 1 million in 1500 AD. But propelled by their new wealth, the Jewish population started to increase rapidly and by 1939 had reached 16.5 million globally.

From an evolutionary perspective, the population increase was the result of a successful adaptation. Because of the requirement for literacy, Jews found themselves better able than non-Jews to cope with the new cognitive demands of urban commerce. "Jews had the behavioral traits conducive to success in a capitalist society," writes the historian Jerry Z. Muller. "They entered commercializing societies with a stock of know-how from their families and communities about how markets work, about calculating profit and loss, about assessing and taking risks. Most important, though hardest to specify, Jews demonstrated a propensity for discovering new wants and to bringing underused resources to market."[14]

Like Chinese immigrant communities, Jews have brought enormous benefits to the economies in which they worked. Unfortunately their success, like that of the immigrant Chinese, has in many cases elicited not gratitude but envy, followed by discrimination or murderous reprisals, a response that reflects more strongly on the greed than the intelligence of their host populations.

From a glance at an Eskimo's physique, it is easy to recognize an evolutionary process at work that has molded the human form for better survival in an arctic environment. Populations that live at high altitudes, like Tibetans, represent another adaptation to extreme environments; in this case, the changes in blood cell regulation are less visible but have been identified genetically. The adaptation of Jews to capitalism is another such evolutionary process, though harder to recognize because the niche to which Jews are adapted is one that has required a behavioral change, not a physical one.

Because of this adaptation, the Jewish population includes proportionately more individuals of higher cognitive capacity than do most others. It thus punches above its weight in endeavors requiring high intelligence. Traits like intelligence are distributed in the shape of a bell, with large numbers of people having the average value and progressively fewer as one moves toward either the higher or lower extreme. It takes only a slight upward shift in the average value to yield significantly more at the upper extreme. Average northern European IQ is 100, by definition, and 4 people per 1,000 in such a population would be expected to have IQs above 140 points. But among Ashkenazim, if the average IQ is taken as 110, then 23 Ashkenazim per 1,000 should exceed 140, the Utah team calculates, a proportion almost six times greater than that in northern Europe. This helps explain why the Jewish population, despite its small size, has produced so many Nobel Prize winners and others of intellectual distinction.

# 9

## THE RISE OF THE WEST

Little by little all the non-western peoples of the earth found it necessary to do something drastic about the intrusive Europeans with their restless, disturbing ways. The rise of the West to this position of dominance all round the globe is, indeed, the main theme of modern world history.

—WILLIAM MCNEILL[1]

Past, present and future, the story of military dynamism in the world is ultimately an investigation into the prowess of Western arms.

—VICTOR DAVIS HANSON[2]

Yet any history of the world's civilizations that underplays the degree of their gradual subordination to the West after 1500 is missing the essential point—the thing most in need of explanation. The rise of the West is, quite simply, the pre-eminent historical phenomenon of the second half of the second millennium after Christ. It is the story at the very heart of modern history. It is perhaps the most challenging riddle historians have to solve.

—NIALL FERGUSON[3]

In 1608 Hans Lippershey, a spectacle maker in the Dutch town of Middelburg, invented the telescope. Within a few decades, telescopes had been introduced from Europe to China, to the Mughal empire in India and to the Ottoman empire. All four civilizations were thus on an equal footing in terms of possessing this powerful new instrument with its latent power for observing the universe and deducing the laws of planetary motion.

There are few controlled experiments in history, but the historian of science Toby Huff has discovered one in the way that the telescope was received and used in the 17th century. The reactions of the four civilizations to this powerful new instrument bear on the very different kinds of society that each had developed.

In Europe the telescope was turned at once toward the heavens. Galileo, hearing a description of Lippershey's device, immediately set to building telescopes of his own. He was first to observe the moons of Jupiter, and he used the fact of Jupiter's satellites as empirical evidence in favor of Copernicus's then disputed notion that the planets, including the Earth, were satellites of the sun. Galileo's argument that the Earth revolved around the sun brought him into conflict with the church's belief that the Earth cannot move. In 1633 he was forced to recant by the Inquisition and placed under house arrest for the rest of his life.

But Europe was not monolithic, and the Inquisition was powerless to suppress the ideas of Copernicus and Galileo in Protestant countries. What Galileo had started was carried forward by Kepler and Newton. The momentum of the Scientific Revolution scarcely faltered.

In the Muslim world, the telescope quickly reached the Mughal empire in India. One was presented in 1616 by the British ambassador to the court of the emperor Jahāngīr, and many more arrived a year later. The Mughals knew a lot about astronomy, but their interest in it was confined to matters of the calendar. A revised calendar was presented to the Mughal emperor Shāh Jahān in 1628, but the scholar who prepared it based it on the Ptolemaic system (which assumes that the sun revolves around an immobile Earth).

Given this extensive familiarity with astronomy, Mughal scholars might have been expected to use the telescope to explore the heavens. But the designers of astronomical instruments in the Mughal empire did not make telescopes, and the scholars created no demand for them. "In the end, no Mughal scholars undertook to use the telescope for astronomical purposes in the seventeenth century," Huff reports.[4]

The telescope fared no better in the other Islamic empire of the time. Telescopes had reached Istanbul by at least 1626 and were quickly incorporated into the Ottoman navy. But despite Muslim eminence in optics in the 14th century, scholars in the Ottoman empire showed no particular interest in the telescope. They were content with the Ptolemaic view of the universe and made no effort to translate the works of Galileo, Copernicus or Kepler. "No new observatories were built, no improved telescopes were manufactured and no cosmological debates about what the telescope revealed in the heavens have been reported," Huff concludes.[5]

Outside of Europe, the most promising new users of the telescope were in China, whose government had a keen interest in astronomy. Moreover, there was an unusual but vigorous mechanism for pumping the new European astronomical discoveries into China in the form of the Jesuit mission there. The Jesuits figured they had a better chance of converting the Chinese to Christianity if they could show that European astronomy provided more accurate calculations of the celestial events in which the Chinese were interested. Through the Jesuits' efforts, the Chinese certainly knew of the telescope by 1626, and the emperor probably received a telescope from Cardinal Borromeo of Milan as early as 1618.

The Jesuits invested significant talent in their mission, which was founded by Matteo Ricci, a trained mathematician who also spoke Chinese. Ricci, who died in 1610, and his successors imported the latest European books on math and astronomy and diligently trained Chinese astronomers, who set about reforming the calendar. One of the Jesuits, Adam Schall von Bell, even became head of the Chinese Bureau of Mathematics and Astronomy.

The Jesuits and their Chinese followers several times arranged prediction challenges between themselves and Chinese astronomers following traditional methods, which the Jesuits always won. The Chinese knew, for instance, that there would be a solar eclipse on June 21, 1629, and the emperor asked both sides to submit the day before their predictions of its exact time and duration. The traditional astronomers predicted the eclipse would start at 10:30 AM and last for two hours. Instead it began at 11:30 AM and lasted two minutes, exactly as the Jesuits had calculated.

But these computational victories did not solve the Jesuits' problem. The Chinese had little curiosity about astronomy itself. Rather, they were interested in divination, in forecasting propitious days for certain events, and astronomy was merely a means to this end. Thus the astronomical bureau was a small unit within the Ministry of Rites. The Jesuits doubted how far they should get into the business of astrological prediction, but their program of converting the Chinese through astronomical excellence compelled them to take the plunge anyway. This led them into confrontation with Chinese officials and to being denounced as foreigners who were interfering in Chinese affairs. In 1661, Schall and the other Jesuits were bound with thick iron chains and thrown into jail. Schall was sentenced to be executed by dismemberment, and only an earthquake that occurred the next day prompted his release.

The puzzle is that throughout this period the Chinese made no improvements on the telescope. Nor did they show any sustained interest in the ferment of European ideas about the theoretical structure of the universe, despite being plied by the Jesuits with the latest

European research. Chinese astronomers had behind them a centuries-old tradition of astronomical observation. But it was embedded in a Chinese cosmological system that they were reluctant to abandon. Their latent xenophobia also supported resistance to new ideas. "It is better to have no good astronomy than to have Westerners in China," wrote the anti-Christian scholar Yang Guangxian.[6]

Both China and the Muslim world suffered from a "deficit of curiosity" about the natural world, Huff says, which he attributes to their educational systems. But the differences between European societies and the others went considerably beyond education and scientific curiosity. The reception of the telescope shows that by the early 17th century, fundamental differences had already emerged in the social behavior of the four civilizations and in the institutions that embodied it. European societies were innovative, outward looking, keen to develop and apply new knowledge, and sufficiently open and pluralistic to prevent the old order from suppressing the new. Those of China and the Islamic world were still entrammeled in traditional religious structures and too subservient to hierarchy to support free-thinking and innovation.

## The Dynamism of the West

The trends in the 17th century illumined by Huff's telescope experiment have continued to the present day with surprisingly little change. Four hundred years later, European societies are still more open and innovative than others. Islamic societies are still inward looking, traditional and hostile to pluralism. An authoritarian regime still rules China, suppresses all organized opposition and inhibits the free flow of ideas and information. The steadiness of these salient trends is a strong indication, though indeed not proof, that evolutionary forces have shaped the basic nature of these societies and their institutions.

Because its societies are open and innovative, the West has come to achieve a surprising degree of dominance in many spheres, despite the fact that its methods and knowledge have long been an open book from which all others may copy. Most of the world is now integrated into the Western economic system. But Japan and China, two of its chief economic rivals, show few signs of becoming better innovators. Most of the world's most successful corporations are still found in the West. Americans and Europeans still dominate most fields of scientific research and collect most Nobel Prizes.

The West continues to lead in military power. Its preeminence has not been uniform. Japan defeated the Russian fleet at the battle of Tsushima in 1905 and seized the East Asian holdings of European colonial powers in the Second World War. China battled the United States to a draw in the Korean War, and the United States did not prevail in Vietnam. European powers have withdrawn from many colonized countries after the cost of holding them became prohibitive. But Western force has remained essentially unchallenged since it withstood the threat of Ottoman invasion in the 17th century. For centuries, the most serious adversaries of Western powers have been other Western nations.

Western science, a driver of technology, still holds a commanding lead over that of other countries. Despite the expectation that Japanese science would become formidable in the wake of its modernization, this flowering has failed to take place. And despite its heavy investment in scientific research, there is no guarantee that by mere force of numbers China can become a major scientific power. Science is essentially subversive in that it requires, at least in its great advances, the toppling of accepted theories and their replacement with better ones. East Asian societies tend to place high value on conformity and respect for superiors, which is not fertile ground for science to flourish.

Throughout the world, Western medicine has proven more effective than the traditional kind. Western music, art and film are generally more creative than the tradition-bound artistic cultures of the East, and the openness of Western societies is regarded by many as more appealing. It would be invalid to ascribe any particular virtue to Europeans as individuals—they are little different from anyone else. But European social organization and especially its institutions have by several significant yardsticks proved more productive and innovative than those of other races. What then explains the rise and continued success of the West?

## Geographic Determinism

One explanation for the rise of the West is geographical. The geographer Jared Diamond is the most recent exponent of the idea. In his well-known book *Guns, Germs, and Steel,* he argues that the West is more powerful than others simply because it got a head start by enjoying more favorable conditions for agriculture. The nature of the people themselves, or their societies, have nothing to do with it, in his view. Everything in human history was determined by geographical features, such as the plant and animal species available for domestication or the plagues endemic in one population but not another.

Despite the popularity of Diamond's book, there are several serious gaps in its argument. One is its anti-evolutionary assumption that only geography matters, not genes. His book, Diamond writes, can be summarized in a single sentence: "History followed different courses for different peoples because of differences among peoples' environments, not because of biological differences among peoples themselves."[7] Geographic determinism, however, is as absurd a position as genetic determinism, given that evolution is about the interaction between the two.

Diamond's book is constructed as an answer to the question he was asked by a New Guinea tribesman as to why Western civilization produced so many more material goods than New Guinean society. Diamond gives no weight to such developments as the rise of modern science, the Industrial Revolution and the economic institutions through which Europeans at last escaped the Malthusian trap. Indeed, when Europeans brought their economic methods to Australia, for instance, they were quickly able to create and operate a European economy. Aborigines, the native Australian population, were still in the Paleolithic Age when Europeans arrived and showed no signs of developing any more advanced material culture.

If in the same environment, that of Australia, one population can operate a highly productive economy and another cannot, surely it cannot be the environment that is decisive, as Diamond asserts, but rather some critical difference in the nature of the two people and their societies.

Diamond himself raises this counterargument, but only to dismiss it as "loathsome" and "racist," a stratagem that spares him the trouble of having to address its merits. While demonizing the opponents of one's views is often effective in the academic sandbox, it is not automatically racist to consider racial categories as a possible explanatory factor. Diamond himself does so when it suits his purpose. He states that "natural selection promoting genes for intelligence has probably been far more ruthless in New Guinea than in more densely populated, politically complex societies. . . . In mental ability, New Guineans are probably genetically superior to Westerners."[8] There is no evidence that this unlikely conjecture is true.[9]

Equally strange is his assertion that intelligence is more likely to be favored in Stone Age societies than in modern ones. Intelligence can be more highly rewarded in modern societies because it is in far greater demand, and the East Asians and Europeans who have built such societies do in fact have higher IQ scores, which may mean higher intelligence, than people who live in tribal or hunter-gatherer societies.

*Guns, Germs, and Steel* has been widely popular, but the many readers who presumably skip over the oddity of its counterfactual statements are missing an important clue to the nature of Diamond's book. It is driven by ideology, not science. The pretty arguments about the availability of domesticable species or the spread of disease are not dispassionate assessments of fact but are harnessed to Diamond's galloping horse of geographic determinism, itself designed to drag the reader away from the idea that genes and evolution might have played any part in recent human history.

Geography and climate have undoubtedly been important, but not to the overwhelming degree that Diamond suggests. The effects of geography are easiest to see in a negative sense, especially their role in holding back population-driven urbanization in regions of low population density, such as Africa and Polynesia. Much harder to understand is how Europe and East Asia, lying on much the same lines of latitude, were driven in the different directions that led to the West's dominance.

If geography provides only a first cut at the answer, can economics provide a more detailed explanation for the rise of the West? As recounted in chapter 7, economic historians have generally looked to factors such as institutions and resources to explain the genesis of the Industrial Revolution. But many of the apparent conditions for success were present in China as well as England, giving little evident reason for the West's preeminence. "Almost every element usually regarded by historians as a major contributory cause to the industrial revolution in north-western Europe was also present in China," concluded the historian Mark Elvin.[10]

Those who favor institutions as the key to the Industrial Revolution have emphasized England's Glorious Revolution of 1688, which placed the sovereign firmly under the control of Parliament and rationalized economic incentives. But both the Glorious Revolution and the Industrial Revolution that followed it were late developments in the rise of the West, the foundations of which historians believe were laid much earlier.

In a recent essay seeking to explain the rise of the West, the historian Niall Ferguson cited six institutions, the first of which he calls competition. By competition he means "a decentralization of political life, which created the launch-pad for both nation-states and capitalism."[11] This is another way of saying that the West, broadly speaking, enjoyed open societies with competing institutions, as opposed to the unrelieved despotism of the East.

The open society made possible the other institutions Ferguson deems critical to the rise of the West, such as the rule of law, including private property rights and the representation of property owners in a legislature; advances in science and medicine; and a growing economy fueled by technology and consumer demand.

"In the course of roughly 500 years," Ferguson writes, "Western civilization rose to a position of extraordinary dominance in the world. . . . Western science shifted the paradigms; others either followed or were left behind. Western systems of law and the political models derived from them, including democracy, displaced or defeated the non-Western alternatives. . . . Above all, the Western model of industrial production and mass consumption left all alternative models of economic organization floundering in its wake."[12]

A society with several different power centers is less likely than an autocracy to suppress new ideas or to thwart innovation and entrepreneurship. Europe thus provided a more favorable environment than China for the emergence of science and medicine, and for the rise of capitalism. But Ferguson's analysis boils down to the assertion that the West succeeded because it was an open society. This is true so far as it goes, but why did the West alone develop a society of this nature? "This openness of society, together with its inventiveness, becomes what is to be explained," writes the economic historian Eric Jones.[13]

## How the West Arose

Some 50,000 years ago, a vast natural experiment was set in motion when modern humans dispersed across the globe from their ancestral African homeland. In Africa, Australasia, East Asia, Europe and the Americas, people developed very different kinds of society depending on the various challenges they faced. For at least the past 500 years, for which detailed records exist, and probably for far longer, these differences have been of an enduring nature.

Nature's experiment, with at least five versions running in parallel for much of the time, has had a complex outcome. What is clear is that from the same human clay, a wide variety of societies can be molded. Australia serves as a kind of baseline. It was inhabited by immigrants from the African homeland some 46,000 years ago. The descendants of these first inhabitants, according to the evidence of their DNA, managed to fend off all outsiders until the arrival of Europeans in the 17th century. At that time, their way of life had changed little. Australian aborigines still lived in tribal societies without towns or cities. Their technology differed little from that of the Paleolithic hunters who reached Europe at the same time their ancestors arrived in Australia. During the 46,000 years of their isolation, they had invented neither the wheel nor the bow and arrow. They lived in a state of perpetual warfare between neighboring tribes. Their most conspicuous cultural achievement was an intense religion, some of whose rituals lasted through day and night for months at a time. The leisure to pursue these elaborate devotions was earned by the aborigines' ability to flourish in a near desert environment in which newcomers would have perished. But for lack of population growth and demographic pressure, aboriginal tribes were never forced into the intense process of state formation and empire building that shaped other civilizations.

In Africa, population numbers were higher than in Australia, agriculture was quickly adopted and settled societies developed. From these gradually emerged more complex societies, including primitive states. But because of low population density, these primitive states did not enter the phase of political rivalry and sustained warfare from which empires emerged in Mesopotamia, the Yellow River valley and, much later, in the Andean highlands. The population of Africa in 1500 was only 46 million. The soil being mostly poor, there were few agricultural surpluses and so no incentive to develop property rights. For lack of the wheel and navigable rivers, transport within Africa was difficult and trade was small scale. For lack of demographic pressure, African societies had little incentive to develop the skills that trade stimulates, to accumulate capital, to develop occupational specialties or to generate modern societies. The phase of state and empire building had only just begun when it was cut short by European colonization.

History in the Americas began only 15,000 years ago, when the first immigrants from Siberia trekked across the then extant land bridge between Siberia and Alaska. Significant empires arose in Mexico, Central America and the Andes. But the populations took many years to attain the critical density for state formation. The Aztecs and Incas had made only a late and uncertain start toward modern states and were already debilitated by internal weaknesses when the conquistadores arrived on their doorstep.

Only in Eurasia did substantial states and empires emerge. More favorable climate and geography allowed larger populations to develop. Under the transforming influences of trade and warfare, empires arose in China, India, the Near East and Europe.

It is hard to identify the influences that may have shaped the European population before about the 5th century AD, when civil authority in the western half of the Roman empire collapsed. In geographical terms, Europe then consisted of a patchwork of cleared regions separated by forests, mountains or marshes. These cleared arable regions became the nucleus of new polities that began to emerge into states around 900 AD. But this defragmentation was a slow process. There were still around 1,000 political units in Europe by the 14th century. Nation-states began to develop in the 15th century. By 1900, Europe consisted of 25 states.[14]

China's geography, by contrast, channeled the social behavior of its population in a very different direction. In the fertile plain between the Yangtze and Yellow rivers, population steadily grew and at an early date was forced into the usual winner-take-all competition between states. China was unified by 221 BC and remained an autocracy, subject to periodic raids from the powerful nomadic peoples along its northern borders.

"Any objective survey of the past 10,000 years of human history," writes the anthropologist Peter Farb, "would show that during almost all of it, northern Europeans were an inferior barbarian race, living in squalor and ignorance, producing few cultural innovations."[15] But during the early Middle Ages, a favorable combination of factors set the stage for Europeans to develop a particularly successful form of social organization. These included a geography that favored the existence of a number of independent states and made it hard for one to dominate all the rest; a population dense enough to encourage social stratification and trade; and an independent center of influence in the form of the church, which set limits on the power of local rulers. By 1200, Europe was still backward compared to

China and the Islamic world, but it had institutions in place that were about to foster an unparalleled burst of innovation accompanied by the rise of science.

# Origins of Modern Science

A distinctive feature of Western civilization is its creation of modern science. By delving into the roots of modern science, might one discover the essential factors that nudged European societies onto their special track?

A careful comparison of early science in Europe, the Islamic world and China has been made by the science historian Toby Huff, whose telescope experiment is recounted above. To anyone who might have surveyed the world in 1200 AD, modern science would have seemed most likely to arise not in Europe but in the Islamic world or in China. The scientific works of ancient Greece were translated into Arabic in the 12th and 13th centuries. People writing in Arabic—who included Jews, Christians and Iranians as well as Arabs—made Arab science the most advanced in the world from the 8th until the 14th century. Scientists writing in Arabic led their fields in mathematics, astronomy, physics, optics and medicine. Arabs perfected trigonometry and spherical geometry.

China too would have seemed fertile ground for science. The three inventions cited by Francis Bacon in 1620 as the greatest known to man—the compass, gunpowder and the printing press—were all Chinese in origin. Besides its technological inventiveness, China had a long history of astronomical observation, a necessary base for understanding the mechanics of the sun's planetary system.

Yet both Arab and Chinese science faltered for essentially similar reasons. Science is not the independent action of lone individuals but a social activity, the work of a community of scholars who check, challenge and build on one another's work. Science therefore needs social institutions, like universities or research institutes, in which to thrive, and these have to be reasonably free of intellectual constraints imposed by religious authorities or government.

In both the Islamic world and China, there proved to be no room for independent institutions. In Islam, there were madrasas, institutes for religious education, attached to mosques. But their prime purpose was to inculcate what were called Islamic sciences, the study of the Qur'an and Islamic law, and not foreign sciences, as the natural sciences were known. Much of ancient Greek philosophy conflicted with Qur'anic teaching and was excluded from study. Scholars who displeased religious authorities could find themselves abruptly silenced by a fatwa. The intellectual tradition of Islam, that the Qur'an and the sayings of Muhammad contained all science and law, created a hostile environment for all independent lines of thought.

Islamic rulers long kept challenges away by forbidding the printing press and squelching troublesome lines of inquiry. In Europe, interest in new knowledge was not confined to an elite but pervaded societies in which literacy was becoming more widespread. By 1500 there were 1,700 printing presses distributed in 300 European cities in every country except Russia.[16] In the Ottoman empire, a decree of Sultan Selim I specified the death penalty for anyone who even used a printing press. Istanbul did not acquire a printing press until 1726 and the owners were allowed to publish only a few titles before being closed down.

Religious authorities in Islamic countries disdained any source of knowledge other than the Qur'an, and frequently exercised their power to suppress it. Institutes like the distinguished Maragha observatory in Iran, founded in 1259 AD, enjoyed only a brief lifetime. As late as 1580, an observatory being built in Istanbul was torn down for religious reasons before it was even completed.[17]

The economist Timur Kuran has recently argued that the Islamic world was held back economically largely because of rigidities in Islamic law regarding commerce. Corporations, for instance, could be dissolved on the death of any partner if his heirs wanted immediate payment. "In sum, several self-enforcing elements of Islamic law—contracting provisions, inheritance system, marriage regulations—jointly contributed to the stagnation of the Middle East's commercial infrastructure," he writes.[18] But it is unconvincing to blame Islamic law; Europeans were faced with similar theologically based laws, such as those against usury, but they made the law accommodate itself to society's larger purposes. In Islam the forces of modernity did not compel the Ottoman state to modernize its legal system until the 19th century.

How is it, then, that Arabic science was so good in the 8th to 14th centuries, despite such inhospitable conditions? The reason, Huff believes, is that in the early centuries of Muslim rule few people had in fact converted to Islam. It was only when the pace of conversion picked up in the 10th century that Muslim majorities became commonplace, a dynamic "which probably had negative consequences for the pursuit of the natural sciences and intellectual life in general."[19]

China, though for different reasons, developed the same antipathy to modern science as did the Islamic world. One problem in China was the absence of any institutions independent of the emperor. There were no universities. Such academies as existed were essentially crammers for the imperial examination system. Independent thinkers were not encouraged. When Hung-wu, the first emperor of the Ming dynasty, decided that scholars had let things get out of hand, he ordered the death penalty for 68 degree holders and 2 students, and penal servitude for 70 degree holders and 12 students. The problem with Chinese science, Huff writes, was not that it was technically flawed, "but that Chinese authorities neither created or tolerated independent institutions of higher learning within which disinterested scholars could pursue their insights."[20] China, unlike the Islamic world, did not ban printing presses, but the books they produced were only for the elite.

Another impediment to independent thought was the stultifying education system, which consisted of rote memorization of the more than 500,000 characters that comprised the Confucian classics, and the ability to write a stylized commentary on them. The imperial examination system, which began in 124 BC, took its final form in 1368 AD and remained unchanged until 1905, deterring intellectual innovation for a further five centuries.

That modern science was for centuries suppressed both in China and in the Islamic world means that its rise in Europe should in no way be taken for granted. Europe too had vested interests resistant to technological change and its attendant disruptions. European religious authorities, just as in Islam, were quick to deter challenges to church doctrine. The bishop of Paris, Étienne Tempier, in 1270 condemned 13 doctrines held by followers of Aristotle, whose philosophy had gained substantial influence in Europe's universities. The bishop followed up in 1277 by prohibiting 219 philosophical and theological theses being discussed at the University of Paris.

But Europe differed from China and the Islamic world in that its educational institutes had considerable independence. The European concept of the corporation as a legal person conferred a certain freedom of thought and action on bodies like guilds and universities. Church authorities could object to what was being taught or discussed, but they could not permanently suppress scientific ideas.

Though Europe's universities started by teaching theology, like the madrasas, they soon moved on to the philosophy of Aristotle, and from philosophy to physics and astronomy. Within these institutions, scientists were able to begin the systematic investigation of nature, thus laying the basis for modern science.

The existence of universities explains how science was able to thrive in Europe though not in China or the Islamic world, but does not explain how science got started in Europe in the first place. What were the preexisting, nonscientific sources from which the scientific enterprise arose?

Huff presents an interesting idea of where to find them. "The riddle of the success of modern science in the West—and its failure in non-Western civilizations—is to be solved by studying the non-scientific domains of culture, that is, religion, philosophy, theology, and the like," he writes.[21]

Christian theology had a rich history of argumentation about fine matters of doctrine, many of them stemming from the complex dogma of the Trinity. These disputes shaped in Europeans' minds the idea of reason as a human attribute. It was reason that separated man from animal. Helped by the rediscovery of Roman civil law toward the end of the 11th century, Europe developed the concept of a legal system. Reason and conscience were adopted as the criteria for deciding legal practice. So it was from there a stone's throw to the concept of laws of nature, to assuming that there existed a Book of Nature and a World Machine that could be comprehended by human reason. It was the revolution in legal thought of the 12th and 13th centuries, in Huff's view, that transformed medieval society in Europe and made it receptive ground for the growth of modern science.

# The Rewards of Openness

The concepts of law and reason in Europe that were the wellsprings of modern science served also as the basis for an open society. Trade and exploration, which Chinese emperors were able to suppress when it suited them, became central forces in Europe's expansion.

Between intermittent bouts of warfare, there was vigorous trade between Europe's various regions. Trade was one of the forces behind the European exploration of the world. The 1490s saw Vasco da Gama's visit to India and Columbus's to the Americas. These voyages also marked a distinctly European curiosity about the world. The exploration was allied with a flood of new technical inventions, the beginnings of modern science and the emergence of capitalism.

It was Europe that discovered the world, not the other way around. The Chinese admiral Zheng He mounted several voyages to Southeast Asia and Africa in the early 15th century, but these were not sustained. Having discovered the rest of the world, Europeans set up trading routes, followed in many instances by conquest. Europeans brushed aside tribal societies almost at will, dispatching settlers to occupy the Americas, Australia and large tracts of Africa.

The roots of European distinctiveness may have been laid as early as the 11th century, if not before, yet even by 1500 Europe's impending rise was far from evident. The Ottoman empire at that time was still expanding. China was enjoying a period of stability under the Ming dynasty. The Mughal empire was about to rise in India. All three powers were more substantial than any in Europe.

Europe lacked the military advantage of being united but could afford its fragmentation, though only narrowly, because, unlike China, it was not under continual threat of invasion. Lying at the western extremity of the Eurasian landmass, Europe was protected on its eastern flank by the buffer states of Russia and Byzantium. From the 10th century on, after onslaughts of Vikings, Magyars and Muslims had been turned back, Europe was reasonably free from external attack, and England, with the extra defense of being an island, enjoyed the greatest security of all.

Hence, unlike the Chinese, Europeans were never forced to seek or accept an autocratic regime strong enough to protect them from outsiders. They had the luxury of preferring independence and of fighting just among themselves. These internal wars let them benefit from the spur of military competition, but the geography and politics of Europe blocked the usual endgame leading to a single permanent empire. The post-Roman empires that arose in Europe, whether of Charlemagne, the Hapsburgs, Napoleon or Hitler, were never complete and tended to be short-lived.

In authoritarian societies, the ruler can coerce taxes, raise armies and wage war. In principle, the authoritarian states of China and the Islamic world should have enjoyed greater military power than Europe's handful of disunited states, each with a sovereign obliged to various degrees to acknowledge local laws and elites. And so for many centuries they did. Europe in the 13th century was no match for the western Mongol army that invaded Poland, Hungary and the Holy Roman Empire with orders to push to the Atlantic coast; only because the Great Khan Ögedei died in 1241, precipitating a succession crisis, did the Mongols voluntarily withdraw from Europe. After the Byzantine state collapsed in 1453, removing the buffer that had separated Europe from the Turkish horde, Ottoman armies were able to penetrate Europe as far as Vienna in 1529 and again in 1683.

But Europe's growing wealth and inventiveness eventually reversed its position of military weakness. Its backwardness in 1500, compared with the Islamic and Chinese empires, was only apparent. European expeditions were soon to conquer India, North and South America, Australia and most of Africa. Europe occupies 7% of the earth's landmass but came to rule 35% of it by 1800 and 84% by 1914.

Unlike in Europe, where science, technology and industry were closely intertwined, technology in China was never harnessed to industry, and industry was never allowed the space for autonomous development. China's enthusiasm for invention had long since ossified. The mandarins had a distaste for novelty. They spurned foreign inventions and lacked the curiosity that drove the intellectually adventurous Europeans to reach beyond technology to the scientific principles behind it.

There was no free market nor institutionalized property rights in China. "The Chinese state was always interfering with private enterprise—taking over lucrative activities, prohibiting others, manipulating prices, exacting bribes, curtailing private enrichment," writes the economic historian David Landes. "Bad government strangled initiative, increased the cost of transactions, diverted talent from commerce and industry."[22]

In the lapidary words of Adam Smith, "Little else is requisite to carry a state to the highest degree of opulence from the lowest barbarism, but peace, easy taxes, and a tolerable administration of justice: all the rest being brought about by the natural course of things."[23] But the "little else" is something of an understatement. Peace, easy taxes and justice are seldom found together in history. Only in Europe was this magic formula achieved, and it became the basis for Europe's unexpected ascent in the world.

# The Adaptive Response to Different Societies

In his book *The Wealth and Poverty of Nations,* the economic historian David Landes examines every possible factor for explaining the rise of the West and the stagnation of China and concludes, in essence, that the answer lies in the nature of the people. Landes attributes the decisive factor to culture, but describes culture in such a way as to imply race.

"If we learn anything from the history of economic development, it is that culture makes all the difference," he writes. "Witness the enterprise of expatriate minorities—the Chinese in East and Southeast Asia, Indians in East Africa, Lebanese in West Africa, Jews and Calvinists throughout much of Europe, and on and on. Yet culture, in the sense of the inner values and attitudes that guide a population, frightens scholars. It has a sulfuric odor of race and inheritance, an air of immutability."[24]

Sulfuric odor or not, the culture of each race, whether genetically based or otherwise, is what Landes suggests has made the difference in economic development. Given the distinctiveness of European societies and the period for which they have been on their own path of development—at least 1,000 years—it is highly likely that the social behavior of Europeans has been adapting genetically to the challenges of surviving and prospering in a European society. The data gathered by Clark on declining rates of violence and increasing rates of literacy from 1200 to 1800, described in chapter 7, are evidence that this is indeed the case.

Though equivalent data does not exist for the Chinese population, their society has been distinctive for even longer—at least 2,000 years—and the intense pressures on survival discussed in chapter 7 would have adapted the Chinese to their society just as Europeans became adapted to theirs.

Psychologists who study the behaviors characteristic of European and East Asian populations usually ascribe everything solely to culture. From an evolutionary perspective, this is implausible. A society's social behavior is central to its survival. Social behavior would have been as closely tailored to prevailing conditions as are the observable features of difference among races such as skin or hair color.

The institutions that characterize a society are a mix of culturally determined and genetically influenced behaviors. The cultural component can be recognized because it has a generally higher rate of change, despite the conservatism of many cultural institutions. Warfare, for instance, is an institution of all human societies, but whether this genetically shaped propensity is exercised depends on culture and circumstances. Germany and Japan developed highly militaristic societies before and during the Second World War but both are now determinedly pacific. This is a cultural change, one far too quick to be genetic. There can be little doubt that both nations retain the propensity for warfare and would exercise it if they needed to do so.

A distinctive feature of genetically shaped behaviors is that they persist unchanged over many generations. The presence of a genetic anchor would explain why expatriate English populations throughout the world have behaved like one another and like their source population over many centuries, and why the same is true of the Chinese abroad. A genetic basis for these groups' social behavior also explains why it is so hard for other populations to copy their desirable features. The Malay, Thai or Indonesian populations who have prosperous Chinese populations in their midst might envy the Chinese success but are strangely unable to copy it. People are highly imitative, and if Chinese business success were purely cultural, everyone would find it easy to adopt the same methods. This is not the case because social behavior, of Chinese and others, is genetically shaped.

The genetic basis of human social behavior is still largely opaque, and it's hard to tell exactly how the neural rules that influence behavior are written. There is clearly a genetic propensity to avoid incest, for example. But it's very unlikely that the genetic rule is written in exactly those terms. Marriage records from Israeli kibbutzim and Chinese families in Taiwan suggest that in practice the incest taboo is driven by an aversion to marrying partners whom one knew intimately in childhood. So the neural rule is probably something like "If you grew up under the same roof with this person, they are not a suitable marriage partner."

Do Europeans carry genes that favor open societies and the rule of law? Is there a gene for respecting property rights or restraining the absolutism of rulers? Obviously this is unlikely to be the case. No one can yet say exactly what patterns in the neural circuitry predispose European populations to prefer open societies and the rule of law to autocracies, or Chinese to be drawn to a system of family obligations, political hierarchy and conformity. But there is no reason to doubt that evolution is capable of framing subtle solutions to complex problems of social adaptation.

There is almost certainly a genetic propensity for following society's rules and punishing those who violate them, as noted in chapter 3. If Europeans were slightly less inclined to punish violators and Chinese more so, that could explain why European societies are more tolerant of dissenters and innovators, and Chinese societies less so. Because the genes that govern rule following and punishment of violators have not yet been identified, it is not yet known if these do in fact vary in European and Chinese populations in the way suggested. Nature has many dials to twist in setting the intensities of the various human social behaviors and many different ways of arriving at the same solution.

The rise of the West was not some cultural accident. It was the direct result of the evolution of European populations as they adapted to the geographic and military conditions of their particular ecological habitat. That European societies have turned out to be more innovative and productive than others, at least under present circumstances, does not of course mean that Europeans are superior to others—a meaningless term in any case from the evolutionary perspective. Europeans are much like everyone else except for minor differences in their social behavior. But these minor differences, for the most part invisible in an individual, have major consequences at the level of a society. European institutions, a blend of both culture and European adaptive social behavior, are the reason that Europeans have constructed innovative, open and productive societies. The rise of the West is an event not just in history but also in human evolution.

# 10

---

# EVOLUTIONARY
# PERSPECTIVES
# ON RACE

Imagine that you, if an English speaker of European descent, are standing on a hill with someone from East Asia and another from Africa. Through a slip in the space-time continuum, you suddenly find that you are holding your mother's hand, and she your grandmother's, and so on through a long line of ancestors that stretches down the hill. The same living ancestors have appeared beside the East Asian and the African, and the three lines of women holding hands snake down the hillside to the valley below.

You let go of your mother's hand and walk down the hill to review the three lineages. The women holding one another's hands are standing 3 feet apart. The average generation time through most of history has been around 25 years, meaning there have been four generations per century. So every 12 feet you walk encompasses a century of ancestresses, and every 120 feet a thousand years.

You pass by your ancestors in wonder but cannot communicate

with them; the shifting languages they speak are now far ancestral to English. Their faces soon lose their distinctively European features, although their skin is still pale. After you have walked 3,600 feet, just over two thirds of a mile, a strange thing happens. A woman is standing between your line of ancestresses and those of the East Asian and at her position the two lines merge into one. She is holding in one hand the hands of her two daughters, one of whom is first in the European line and the other the first in the East Asian line.

As you continue down the hill, you are reviewing just two lineages, the now joint European–East Asian line and that of Africans. The people in the joint line grow steadily darker in complexion, since they lived before humans expanded to extreme northern latitudes and developed pale skin. Then, after you've been walking just over a mile, it is the turn of these two lineages to converge into one. There stands a woman holding the hands of two daughters, one of whom stayed in Africa and the other joined the small hunter-gatherer band that left the ancestral homeland some 50,000 years ago. In a walk of some 22 minutes, the human species has been reunified before your eyes.

Had you continued walking for another hour, all of it along African ancestors, you would have reached the 200,000 year mark, the earliest known appearance of modern humans. Three quarters of modern human existence has been spent in Africa, only the last quarter outside it. Today's races hold three quarters of their history in common, only one quarter apart.[1]

From an evolutionary perspective, the human races are all very similar variations of the same gene pool. The question that looms over all the social sciences, unanswered and largely unaddressed, is how to explain the paradox that people as individuals are so similar yet human societies differ so conspicuously in their cultural and economic attainments.

The argument presented in the pages above is that these differences do not spring from any great disparity between the individual members of the various races. Rather, they stem from the quite minor variations in human social behavior, whether of trust, conformity, aggressiveness or other traits, that have evolved within each race during its geographical and historical experience. These variations have set the framework for social institutions of significantly different character. It is because of their institutions—which are largely cultural edifices resting on a base of genetically shaped social behaviors—that the societies of the West and of East Asia are so different, that tribal societies are so unlike modern states, and that rich countries are rich and poor countries deprived.

The consensus explanation of almost all social scientists is that human societies differ only in their culture, with the implicit premise that evolution has played no role in the differences between populations. But the all-culture explanation is implausible for several reasons.

First, it is of course a conjecture. No one can at present say what precise mix of genetics and culture underlies the differences between human societies, and the assertion that evolution plays no role is merely a surmise.

Second, the all-culture position was formulated largely by the anthropologist Franz Boas as an antiracist position, which may be laudable in motive, but political ideology of any kind has no proper place in science. Moreover Boas wrote at a time before it was known that human evolution had not halted in the distant past.

Third, the all-culture conjecture does not satisfactorily explain why the differences between human societies are as deep-rooted as seems to be the case. If the differences between a tribal society and a modern state were purely cultural, it should be easy to modernize a tribal society by importing Western institutions. American experience in Haiti, Iraq and Afghanistan generally suggests otherwise. Culture undeniably explains many important differences between societies. The issue is whether it is a sufficient explanation for all such differences.

Fourth, the all-culture conjecture is severely lacking in proper care and maintenance. Its adherents have failed to update it to take account of the new discovery that human evolution has been recent, copious and regional. Their hypothesis must assume, against all the evidence that has accumulated over the past 30 years, that the mind is a blank slate, born immaculately bereft of any innate behavior, and that the importance of social behavior for survival is too trivial to have been molded by natural selection. Or, if they allow that social behavior does have a genetic basis, they must explain how it could have remained unchanged in all races, despite the vast changes in human social structure over the past 15,000 years, when many other traits are now known to have evolved independently in each race, transforming some 14% of the human genome.

The thesis presented here assumes, to the contrary, that there is a genetic component to human social behavior; that this component, so critical to human survival, is subject to evolutionary change and has indeed evolved over time; that the evolution in social behavior has necessarily proceeded independently in the five major races and others; and that slight evolutionary differences in social behavior under-lie the differences in social institutions prevalent among the major human populations.

Like the all-culture position, this thesis is unproven, but it rests on several premises that are plausible in the light of new knowledge.

The first is that the social structures of primates, humans included, are based on genetically shaped behaviors. Chimpanzees inherited a genetic template for operation of their distinctive societies from their joint ancestor with humans. The joint ancestor would have bequeathed the same template to the human lineage, which then evolved to support the distinctive features of human social structure, from the pair bonding that emerged some 1.7 million years ago to the emergence of hunter-gatherer bands and tribes. It is hard to see why humans, as an intensely social species, should ever have lost the genetic template for the suite of social behaviors on which their society depends, or why the template should not have continued to evolve during the most dramatic of all its transformations, the shift that enabled the size of human societies to expand from a maximum of 150 in the hunter-gatherer group to vast cities teeming with tens of millions of inhabitants. This transformation, it should be noted, had to evolve independently in the major races since it occurred after they split apart.

A variety of data, including experiments with very young children, points to innate social propensities for cooperativeness, helping others, obeying rules, punishing those who don't, trusting others selectively and a sense of fairness. The genes that direct the neural circuitry for such behaviors are for the most part unknown. But it is plausible that they exist, and genetic systems involving the control of the enzyme MAO-A, associated with aggression, and the hormone oxytocin, a modulator of trust, are already known.

A second premise is that these genetically shaped social behaviors undergird the institutions around which human societies are constructed. Given that such behaviors exist, it seems uncontroversial that institutions should depend on them, and the proposition is endorsed by authorities such as the economist Douglass North and the political scientist Francis Fukuyama, both of whom see institutions as founded in the genetics of human behavior.

A third premise is that the evolution of social behavior has continued during the past 50,000 years and throughout the historical period. This phase of evolution has necessarily occurred independently and in parallel in the three major races after they split apart and each made the transition from hunting and gathering to settled life. Evidence in the genome that human evolution has been recent, copious and regional provides general support for this thesis, unless any reason can be shown why social behavior should have been exempt from natural selection.

The best possible proof of the premise would be identification of the genes that shape the neural circuitry for social behaviors, and demonstration that they have been under natural selection in each race. No such test is yet available because the genes that underlie social behavior are largely unknown. But brain genes of unknown duties are among the genes found to have been under recent selective pressure in the three principal races, proving that the genes for neural function are not exempt from recent evolutionary change. In addition, the MAO-A gene, which influences aggressivity, varies substantially among races and ethnicities in a way that suggests, though does not prove, that the gene has been under evolutionary pressure.

A fourth premise is that evolved social behavior can in fact be observed in today's various populations. The behavioral changes documented in the English population during the 600 years that preceded the Industrial Revolution include a decline in violence and increases in literacy, the propensity to work and the propensity to save. The same evolutionary shift presumably occurred in the other agrarian populations of Europe and East Asia before they entered their own industrial revolutions. Another behavioral change is evident in the Jewish population as it adapted over the centuries first to educational demands and then to exacting professional niches.

A fifth premise is that the significant differences are those between human societies, not their individual members. Human nature is essentially the same worldwide. But minor variations in social behavior, though barely perceptible, if at all, in an individual, combine to create societies of very different character. These evolutionary differences between societies on the various continents may underlie major and otherwise imperfectly explained turning points in history such as the rise of the West and the decline of the Islamic world and China, as well as the economic disparities that began to emerge in the past few centuries.

To assert that evolution has played some role in human history does not mean that that role is necessarily prominent, let alone decisive. Culture is a mighty force, and people are not slaves to innate propensities, which in any case only prompt the mind in a certain direction. But if all individuals in a society have similar propensities, however slight, toward greater or less social trust, say, or greater or lesser conformity, then the society will tend to act in that direction and to differ from societies that lack such propensities.

# History as If Evolution Mattered

How might historians write if they believed that evolution were relevant to their concerns? They would surely pay greater attention to the evolutionary role of forces like demography and warfare in shaping human societies. Population growth seems to have been the driving force that compelled societies to devise more complex structures, both in order to organize larger numbers of people and for defense against neighbors who were also expanding in numbers and territory. Under the pressure of war, chiefdoms coalesced into archaic states and states into empires. But this sanguinary process faltered if populations were too sparse or people could escape elsewhere.

The forces of natural selection that work within a society have been equally significant. Agrarian economies have kept people striving at the edge of starvation for millennia, the condition in which Darwin perceived that natural selection would favor even the slightest survival advantage. Under these Malthusian conditions, the ratchet of wealth—the ability of the rich to raise more surviving children—slowly diffused the social behaviors required for modern prosperity into the wider society.

These forces have worked independently on the populations in each continent, driving them along paths that were parallel to a large extent but ultimately diverged. Early states arose in East Asia, Europe, Africa and the two Americas. In Australia, however, population numbers and climate remained too adverse to induce the development of agriculture or social structures more elaborate than that of hunter-gatherers.

Human societies of distinctive character arose on all five continents and some became the basis for major civilizations. Historians reject thinking in racial categories for understandable reasons. But it is an error to exclude any possible role for evolution in history. The major civilizations occur within the two main races of East Asians and Caucasians, as distinguished by genetics. Within the East Asian race arose the civilizations of China, Korea and Japan, as well as Siberian steppe cultures such as the Mongols. Within the Caucasian group are the civilizations of India, Russia, the West, South America and the Islamic world.

A primary effect of genetics is to add a substantial degree of inertia or stability to the social behavior and hence to the institutions of each society. Rapid change must be due to culture, not genetics, but if the core social behaviors of each civilization have an evolutionary foundation, as argued in the previous chapter, then the rate of change in their relationships is likely to be constrained. The slow march of evolution, in other words, exerts an unseen collar on the pace of history.

This constraint has considerable bearing on issues such as whether the West will continue its dominance or enter into decline. "What we are living through now is the end of 500 years of Western predominance. This time the Eastern challenger is for real," the historian Niall Ferguson wrote in 2011.[2] Ferguson's basic argument is that empires have always risen and fallen, therefore the United States too will be eclipsed, and the most likely successor on the horizon is China. But the rise and fall of civilizations is in fact vastly slower than that of empires. In Europe, the empires of Charlemagne, the

Hapsburgs, Napoleon and Hitler all rose and fell, without having any significant effect on the rise of Western civilization. Dynasties have changed in China, some of them led by invaders like the Mongols or Manchus, without altering the essential character of Chinese social behavior. Empires are an epiphenomenon upon the surface of the stronger, slower tides of evolution.

Of greater moment are the clashes between the world's civilizations. War was the mechanism that welded early human societies into the first primitive states and has been a constant shaper of state organization ever since. There is no clear reason why continued militarism should not have culminated in a single worldwide empire as soon as transport and communications permitted. The Mongol imperium, a rapacious and highly destructive society that stretched from Eastern Europe to the Sea of Japan, was a prototype of such a universal empire. The Mongol sack of Baghdad destroyed the leading center of Islamic culture. The capitals of Europe nearly suffered the same fate: if the Mongol army that conquered Poland and Hungary had continued its march to the Atlantic coast, as was its plan, the rise of the West would have been aborted or at the least substantially delayed.

Western civilization was certainly expansionary, but after a comparatively brief colonial phase it has refocused on the trade and productive investment that drove its expansion in the first place. It seems a fortunate outcome that the world's dominant military power has turned out to be the West, with a system of international trade and law that offers benefits to all participants, and not a purely predatory and militaristic state like that of the Mongols or Ottomans, as might have been expected, or even a civilized but autocratic one like that of China.

From an evolutionary perspective, an imminent decline of the West seems unlikely. Western social behavior, the source of the open society and open economy with their rewards to innovation, has been shaped by evolution as well as by culture and history and is unlikely to change anytime soon. The West was more exploratory and innovative than other civilizations in 1500 and it is the same way now. Neither Japan nor China has yet seriously challenged the West's preeminence in science and technology despite ample investments and a large body of educated and capable scientists. Well-performing institutions don't guarantee the West's permanent dominance but the social behavior that underlies them is an asset that is likely to persist for many generations, barring some major setback. East Asian societies seem too authoritarian and conformist, despite the high abilities of their citizens, to challenge the innovation of the West, a fact implicitly acknowledged in the Chinese state's intense efforts to steal Western technical and commercial secrets.

But the success of the West, even if long lasting, is necessarily provisional. The framework of social behavior at the root of the West's critical institutions may be frailer than it seems and vulnerable to being overwhelmed by adverse cultural forces such as political stasis, class warfare or a failure of social cohesion. Western societies are well adapted to present economic conditions, which they have in large measure created. In different conditions, the West's advantage might disappear. If the present climatic regime should change substantially, for instance in the global cooling that will precede the inevitable onset of the next ice age, more authoritarian societies like those of East Asia could be better positioned to endure harsh stresses. By evolution's criterion of success, East Asians are already the most successful human population: the Han Chinese are the world's most numerous ethnic group. By another biological criterion, the population of Africa is the most important, since it harbors the most genetic diversity and hence a larger share of the human genetic patrimony than any other race.

The various races and ethnicities into which humans have evolved represent a grand experiment in which nature has tested out some of the variations inherent in the human genome. The experiment is not being conducted in our interests—it has no purpose or goal—yet it offers considerable benefits. Instead of there being a single type of human society, there are many, creating a rich diversity of cultures whose more promising features can be adopted and improved on by others. Without Western production efficiencies, the countries of East Asia might still be locked in stagnant autocracies. Within the West, the success of Jews has benefited every economy in which they worked and contributed immeasurably to the arts and sciences. The strong cultures of East Asia may yet find ways to surpass the West, as they have done for most of their previous history.

## Understanding Race

The idea that human populations are genetically different from one another has been actively ignored by academics and policy makers for fear that such inquiry might promote racism. The argument offered here is that people the world over are highly similar as individuals but that societies differ widely because of evolutionary differences in social behavior. It would be better to take account of evolutionary differences than to continue to ignore them.

Moreover, fears that the evolutionary understanding of race will promote a new phase of racism or imperialism are surely exaggerated. The lessons of past abuses are still vivid enough. Science may be an autonomous body of knowledge, but its interpretation depends strongly on the intellectual climate of the time. In the 19th century, a period of vigorous European expansion, people looked to Social Darwinism to justify dominion over others and deny welfare to the poor. This interpretation of Darwinism has been so thoroughly repudiated that it is hard to conceive of any circumstance in which it could be successfully resurrected.

But is it not a form of racism to link the success of the West to the genetics of Westerners? For several reasons, this is not the case. First, there is no assertion of superiority, which is the essence of racism, and in any case the success of the West is provisional. Its economies are an open book, free for all others to copy, as they are doing, and to improve on. As everyone understands, China is a rising power whose role in the world has yet to be defined. Nations are compared on metrics such as economic or military power, which are constantly shifting and allow none the right or reason to claim permanent dominance, let alone inherent superiority.

Second, a society's achievements, whether in economics or the arts or military preparedness, rests in the first place on its institutions, which are largely cultural in essence. Genes may nudge social behavior in one direction or another, thus affecting the nature of a society's institutions on the timescale of the generations and setting the framework within which culture operates, but this is a long-term effect that leaves ample room for culture to play a major role.

Third, all human races are variations on a common theme. There is no basis from an evolutionary perspective, or any other, for declaring any one variation superior to any other.

One reason why discussion of genetics is so fraught is because of the assumption that genes are immutable and that to say a person or group of people carries a disadvantageous gene puts them beyond remedy. This is at best a partial truth.

The genes whose effects cannot be changed, like those that direct the color of skin or hair or the proportions of the body, are or should be of no relevance to the success of a modern economy. The important genes, at least in terms of the differences between civilizations, are those that influence social behavior.

But the genes that govern human behavior seldom issue imperatives. They operate by setting mere inclinations, of which even the strongest can be overridden. There are almost certainly genes that predispose people to regard incest as abhorrent, yet cases of incest are far from rare because those neural prohibitions can be ignored. Because the prompting of behavioral genes can be resisted, ingrained social behavior may be subject to a variety of manipulations, ranging from education and social pressure to tax incentives. In short, many social behaviors are modifiable and this is probably the case even if they are genetically influenced. Where behavior is concerned, genetic does not mean immutable.

Many forms of new knowledge are potentially dangerous, the energy of the atom being a preeminent example. But instead of curtailing inquiry Western societies have in general assumed that the better policy is to continue exploration in confidence that the rewards can be reaped and the risks managed. It is hard to see why exploration of the human genome and its racial variations should be made an exception to this principle, even though researchers and their audience must first develop the words and concepts to discuss a dangerous subject objectively.

Knowledge is usually considered a better basis for policy than ignorance. This book has been an attempt, undoubtedly imperfect, to dispel the fear of racism that overhangs discussion of human group differences and to begin to explore the far-reaching implications of the discovery that human evolution has been recent, copious and regional.

# ACKNOWLEDGMENTS

This book, like its predecessor *The Faith Instinct,* grew out of *Before the Dawn,* an account of human evolution in the past 50,000 years.

*The Faith Instinct* examined the evolutionary role of religion as a cohesive force in human societies. This book explores new data from the human genome that has shed light on the emergence of the various races of humankind. Both religion and race are essential but strangely unexplored aspects of the human experience. Like everything else in biology, they make no sense except in the light of evolution.

I thank Peter Matson of Sterling Lord Literistic for guiding the initial focus of the book. I am most grateful to Scott Moyers of Penguin Press for thoroughly critiquing the book and guiding it past many perilous shoals with unswerving editorial skill.

I owe a great debt to friends who read early drafts and saved me from many errors of fact and judgment, including Nicholas W. Fisher of the University of Aberdeen, Jeremy J. Stone of Catalytic Diplomacy, Richard L. Tapper of the London School of Oriental and African Studies and my son Alexander Wade of Doctors Without Borders.

# NOTES

CHAPTER 1: EVOLUTION, RACE AND HISTORY

1. Joshua M. Akey, "Constructing Genomic Maps of Positive Selection in Humans: Where Do We Go from Here?" *Genome Research* 19 (2009): 711–22.
2. Xin Yi et al., "Sequencing of 50 Human Exomes Reveals Adaptation to High Altitude," *Science* 329, no. 5987 (July 2, 2010): 75–78.
3. Emmanuel Milot et al., "Evidence for Evolution in Response to Natural Selection in a Contemporary Human Population," *Proceedings of the National Academy of Sciences* 108 (2011): 17040–45.
4. Stephen C. Stearns et al., "Measuring Selection in Contemporary Human Populations," *Nature Reviews Genetics* 11, no. 9 (Sept. 2010): 1–13.
5. American Anthropological Association, "Race: A Public Education Project," www.aaanet.org/resources/A-Public-Education-Program.cfm.
6. Alan H. Goodman, Yolanda T. Moses, and Joseph L. Jones, *Race: Are We So Different?* (Arlington, VA: American Anthropological Association 2012), 2.
7. American Sociological Association, "The Importance of Collecting Data and Doing Social Scientific Research on Race," (Washington, DC: American Sociological Association, 2003), www2.asanet.org/media/asa_race_statement.pdf.
8. Christopher F. Chabris et al., "Most Reported Genetic Associations with General Intelligence Are Probably False Positives," *Psychological Science* 20, no. 10 (Sept. 24, 2012): 1–10.
9. David Epstein, *The Sports Gene: Inside the Science of Extraordinary Athletic Performance* (New York: Current, 2013), 176.

CHAPTER 2: PERVERSIONS OF SCIENCE

1. Richard Hofstadter, *Social Darwinism in American Thought* (Boston: Beacon Press, 1992), 171.
2. Benjamin Isaac, *The Invention of Racism in Classical Antiquity* (Princeton, NJ: Princeton University Press, 2004), 23.
3. Nell Irving Painter, "Why White People Are Called 'Caucasian'?" paper presented at the Fifth Annual Gilder Lehrman Center International Conference, Yale University, New Haven, CT, Nov. 7–8, 2003, www.yale.edu/glc/events/race/Painter.pdf.
4. Jason E. Lewis et al., "The Mismeasure of Science: Stephen Jay Gould Versus Samuel George Morton on Skulls and Bias," *PLoS Biology* 9, no. 6 (June 7, 2011), www.plosbiology.org/article/info%3Adoi%2F10.1371%2Fjournal.pbio.1001071.
5. Hofstadter, *Social Darwinism*, xvi.
6. Charles Darwin, *The Descent of Man and Selection in Relation to Sex*, 2d ed. (New York: Appleton, 1898), 136.
7. Nicholas Wright Gillham, *A Life of Sir Francis Galton: From African Exploration to the Birth of Eugenics* (New York: Oxford University Press, 2001), 166.
8. Ibid., 357.
9. Edwin Black, *War Against the Weak: Eugenics and America's Campaign to Create a Master Race* (New York: Four Walls Eight Windows, 2003), 37.
10. Ibid., 45–47.
11. Ibid., 90.
12. Daniel J. Kevles, *In the Name of Eugenics: Genetics and the Uses of Human Heredity* (New York: Knopf, 1985), 69.
13. Black, *War Against the Weak*, 87.
14. Ibid., 99.
15. Kevles, *In the Name of Eugenics*, 81.
16. Ibid., 106.
17. Black, *War Against the Weak*, 123.
18. Kevles, *In the Name of Eugenics*, 97.
19. Black, *War Against the Weak*, 393.
20. Madison Grant, *The Passing of the Great Race; or, The Racial Basis of European History*, 4th ed. (New York: Charles Scribner, 1932), 170.
21. Ibid., 263.
22. Jonathan P. Spiro, *Defending the Master Race: Conservation, Eugenics and the Legacy of Madison Grant* (Burlington: University of Vermont Press, 2009), 375.
23. Black, *War Against the Weak*, 100.
24. Ibid., 259.

25. Kevles, *In the Name of Eugenics,* 117.
26. Ibid., 118.
27. Raul Hilberg, *The Destruction of the European Jews* (New York: Holmes & Meier, 1985, student edition), 31.
28. Yvonne Sherratt, *Hitler's Philosophers* (New Haven, CT: Yale University Press, 2013) 60.

CHAPTER 3: ORIGINS OF HUMAN SOCIAL NATURE

1. Bernard Chapais, *Primeval Kinship: How Pair-Bonding Gave Birth to Human Society* (Cambridge, MA: Harvard University Press, 2008), 4.
2. Charles Darwin, *The Descent of Man and Selection in Relation to Sex,* 2d ed. (New York: Appleton, 1898), 131.
3. Michael Tomasello, *Why We Cooperate* (Cambridge, MA: MIT Press, 2009), 27.
4. Ibid., 23.
5. Ibid., 7.
6. Esther Herrmann et al., "Humans Have Evolved Specialized Skills of Social Cognition: The Cultural Intelligence Hypothesis," *Science* 317, no. 5843 (Sept. 7, 2007): 1360–66.
7. Michael Tomasello and Malinda Carpenter, "Shared Intentionality," *Developmental Science* 10, no. 1 (2007): 121–25.
8. Cade McCall and Tania Singer, "The Animal and Human Neuroendocrinology of Social Cognition, Motivation and Behavior," *Nature Neuroscience* 15 (2012): 681–88.
9. Carsten K. W. De Dreu et al., "Oxytocin Promotes Human Ethnocentrism," *Proceedings of the National Academy of Sciences* 108, no. 4 (Jan. 25, 2011), 1262–66.
10. David H. Skuse et al., "Common Polymorphism in the Oxytocin Receptor Gene (OXTR) Is Associated With Human Recognition Skills," *Proceedings of the U.S. National Academy of Sciences* (December 23, 2013).
11. Reviewed in Zoe R. Donaldson and Larry J. Young, "Oxytocin, Vasopressin and the Neurogenesis of Sociality," *Science* 322, no. 5903 (Nov. 7, 2008): 900–904.
12. Nicholas Wade, "Nice Rats, Nasty Rats: Maybe It's All in the Genes," *New York Times,* July 25, 2006, www.nytimes.com/2006/07/25/health/25rats.html?pagewanted=all&_r=0 (accessed Sept. 25, 2013).
13. Robert R. H. Anholt and Trudy F. C. Mackay, "Genetics of Aggression," *Annual Reviews of Genetics* 46 (2012): 145–64.
14. Guang Guo et al., "The VNTR 2 Repeat in MAOA and Delinquent Behavior in Adolescence and Young Adulthood: Associations and MAOA Promoter Activity," *European Journal of Human Genetics* 16 (2008): 624–34.

15. Yoav Gilad et al., "Evidence for Positive Selection and Population Structure at the Human MAO-A Gene," *Proceedings of the National Academy of Sciences* 99, no. 2 (Jan. 22, 2002): 862–67.

16. Kevin M. Beaver et al., "Exploring the Association Between the 2-Repeat Allele of the MAOA Gene Promoter Polymorphism and Psychopathic Personality Traits, Arrests, Incarceration, and Lifetime Antisocial Behavior," *Personality and Individual Differences* 54, no. 2 (Jan. 2013): 164–68.

17. Laura Bevilacqua et al., "A Population-Specific HTR2B Stop Codon Predisposes to Severe Impulsivity," *Nature* 468, no. 7327 (Dec. 23, 2010): 1061–66.

18. Edward O. Wilson, *Sociobiology: The New Synthesis* (Cambridge, MA: Harvard University Press, 1975), 547–75.

19. Edward O. Wilson, *On Human Nature* (Cambridge, MA: Harvard University Press, 1978), 167.

20. Sarah A. Tishkoff et al., "Convergent Adaptation of Human Lactase Persistence in Africa and Europe," *Nature Genetics* 39, no. 1 (Jan. 2007): 31–40.

21. Hillard S. Kaplan, Paul L. Hooper, and Michael Gurven, "The Evolutionary and Sociological Roots of Human Social Organization," *Philosophical Transactions of the Royal Society B: Biological Science* 364, no. 1533 (Nov. 12, 2009): 3289–99.

CHAPTER 4: THE HUMAN EXPERIMENT

1. Charles Darwin, *The Descent of Man and Selection in Relation to Sex,* 2d ed. (New York: Appleton, 1898), 171.

2. Ian Tattersall and Rob DeSalle, *Race? Debunking a Scientific Myth* (College Station: Texas A&M University Press, 2011).

3. J. Craig Venter, *A Life Decoded: My Genome, My Life* (New York: Penguin Books, 2008).

4. Jared Diamond, "Race Without Color," *Discover,* Nov. 1994.

5. Francis S. Collins and Monique K. Mansoura, "The Human Genome Project: Revealing the Shared Inheritance of All Humankind," *Cancer* supplement, Jan. 2001.

6. Jerry A. Coyne, "Are There Human Races?" *Why Evolution Is True,* http://whyevolutionistrue.wordpress.com/2012/02/28/are-there-human-races.

7. Ashley Montagu, *Man's Most Dangerous Myth: The Fallacy of Race,* 6th ed. (Lanham, MD: AltaMira Press/Rowman & Littlefield, 1997), 41.

8. Ibid., 47.

9. Norman J. Sauer, "Forensic Anthropology and the Concept of Race: If Races Don't Exist, Why Are Forensic Anthropologists So Good at Identifying Them?" *Social Science and Medicine* 34, no. 2 (Jan. 1992): 107–11.

10. Winthrop D. Jordan, *The White Man's Burden: Historical Origins of Racism in the United States* (New York: Oxford University Press, 1974), xi–xii.
11. John Novembre et al., "Genes Mirror Geography Within Europe," *Nature* 456, no. 7218 (Nov. 6, 2008): 98–101.
12. Colm O'Dushlaine et al., "Genes Predict Village of Origin in Rural Europe," *European Journal of Human Genetics* 18, no. 11 (Nov. 2010): 1269–70.
13. See, for example, *The History and Geography of Human Genes* (Princeton, NJ: Princeton University Press, 1994), a classic work by L. Luca Cavalli-Sforza, Paolo Menozzi, and Alberto Piazza.
14. Esteban J. Parra, "Human Pigmentation Variation: Evolution, Genetic Basis, and Implications for Public Health," *Yearbook of Physical Anthropology* 50 (2007): 85–105.
15. Rebecca L. Lamason et al., "SLC24A5, a Putative Cation Exchanger, Affects Pigmentation in Zebrafish and Humans," *Science* 310, no. 5755 (Dec. 16, 2005): 1782–86.
16. Akihiro Fujimoto et al., "A Scan for Genetic Determinants of Human Hair Morphology: *EDAR* Is Associated with Asian Hair Thickness," *Human Molecular Genetics* 17, no. 6 (Mar. 15, 2008): 835–43.
17. Yana G. Kamberov et al., "Modeling Recent Human Evolution in Mice by Expression of a Selected EDAR Variant," *Cell* 152, no. 4 (Feb. 14, 2013): 691–702.
18. Koh-ichiro Yoshiura et al., "A SNP in the ABCC11 Gene Is the Determinant of Human Earwax Type," *Nature Genetics* 38, no. 3 (Mar. 2006): 324–30.

CHAPTER 5: THE GENETICS OF RACE

1. Charles Darwin, *The Descent of Man and Selection in Relation to Sex,* 2d ed. (New York: Appleton, 1898), 132.
2. A. M. Bowcock et al., "High Resolution of Human Evolutionary Trees with Polymorphic Microsatellites," *Nature* 368, no. 6470 (Mar. 31, 1994): 455–57.
3. Neil Risch, Esteban Burchard, Elad Ziv, and Hua Tang, "Categorization of Humans in Biomedical Research: Genes, Race and Disease," *Genome Biology* 3, no. 7 (March 2002), http://genomebiology.com/2002/3/7/comment/2007.
4. Noah A. Rosenberg et al., "Genetic Structure of Human Populations," *Science* 298, no. 5602 (Dec. 20, 2002): 2381–85.
5. Frank B. Livingstone and Theodosius Dobzhansky, "On the Non-Existence of Human Races," *Current Anthropology* 3 no. 3 (June 1962): 279.
6. David Serre and Svante Pääbo, "Evidence for Gradients of Human Genetic Diversity Within and Among Continents," *Genome Research* 14 (2004): 1679–85.

7. Noah A. Rosenberg et al., "Clines, Clusters, and the Effect of Study Design on the Inference of Human Population Structure," *PLoS Genetics* 1, no. 6 (2005): 660–71.

8. Jun Z. Li et al., "Worldwide Human Relationships Inferred from Genome-Wide Patterns of Variation," *Science* 319, no. 5866 (Feb. 22, 2008): 1100–1104.

9. Sarah A. Tishkoff et al., "The Genetic Structure and History of Africans and African Americans," *Science* 324, no. 5930 (May 22, 2009): 1035–44.

10. Benjamin F. Voight, Sridhar Kudaravalli, Xiaoquan Wen, Jonathan K. Pritchard, "A Map of Recent Positive Selection in the Human Genome," *PLoS Biology* 4, no. 3 (Mar. 2006): 446–53.

11. Sharon R. Grossman et al., "Identifying Recent Adaptations in Large-Scale Genomic Data," *Cell* 152, no. 4 (Feb. 14, 2013): 703–13.

12. Ibid. These figures are not provided in the text but can be gleaned from a supplementary spreadsheet, Table S2.

13. Joshua M. Akey, "Constructing Genomic Maps of Positive Selection in Humans: Where Do We Go from Here?" *Genome Research* 19, no. 5 (May 2009): 711–22.

14. Ralf Kittler, Manfred Kayser, and Mark Stoneking, "Molecular Evolution of *Pediculus humanus* and the Origin of Clothing," *Current Biology* 13, no. 16 (Aug. 19, 2003): 1414–17. Another louse researcher, David Reed, has argued that a much older date, perhaps 500,000 years ago, is correct.

15. David López Herráez et al., "Genetic Variation and Recent Positive Selection in Worldwide Human Populations: Evidence from Nearly 1 Million SNPs," *PLoS One* 4, no. 11 (Nov. 18, 2009): 1–16.

16. Graham Coop et al., "The Role of Geography in Human Adaptation," *PLoS Genetics* 5, no. 6 (June 2009): 1–16.

17. Matthew B. Gross and Cassandra Kniffen, "Duffy Antigen Receptor for Chemokines: DARC," *Online Mendelian Inheritance in Man,* Dec. 10, 2012, http://omim.org/entry/613665.

18. C. T. Miller et al., "cis-Regulatory Changes in Kit Ligand Expression and Parallel Evolution of Pigmentation in Sticklebacks and Humans," *Cell* 131 (2007): 1179–89.

19. Ryan D. Hernandez et al., "Classic Selective Sweeps Were Rare in Recent Human Evolution," *Science* 331, no. 6019 (Feb. 18, 2011): 920–24.

20. Jonathan K. Pritchard, "Adaptation—Not by Sweeps Alone," *Nature Reviews Genetics* 11, no. 10 (Oct. 2010): 665–67.

21. Hua Tang et al., "Genetic Structure, Self-Identified Race/Ethnicity, and Confounding in Case-Control Association Studies," *American Journal of Human Genetics* 76, no. 2 (Feb. 2005): 268–75.

22. Roman Kosoy et al., "Ancestry Informative Marker Sets for Determining Continental Origin and Admixture Proportions in Common Populations in America," *Human Mutation* 30, no. 1 (Jan. 2009): 69–78.

23. Wenfei Jin et al., "A Genome-Wide Detection of Natural Selection in African Americans Pre- and Post-Admixture," *Genome Research* 22, no. 3 (Mar. 1, 2012): 519–27.

24. Richard Lewontin, "The Apportionment of Human Diversity," *Evolutionary Biology* 6 (1972): 396–97, quoted in Ashley Montagu, *Man's Most Dangerous Myth: The Fallacy of Race,* 6th ed. (Lanham, MD: AltaMira Press/Rowman & Littlefield, 1997), 45–46.

25. Quoted in Daniel L. Hartl and Andrew G. Clark, *Principles of Population Genetics,* 3d ed. (Sunderland, MA: Sinauer Associates, 1997), 119.

26. Quoted by Henry Harpending and Alan R. Rogers, "Genetic Perspectives in Human Origins and Differentiation," *Annual Review of Genomics and Human Genetics* 1 (2000): 361–85.

27. A.W.F. Edwards, "Human Genetic Diversity: Lewontin's Fallacy," *BioEssays* 25, no. 8 (Aug. 2003): 798–801.

28. Ed Hagen, "Biological Aspects of Race," American Association of Physical Anthropologists position statement, *American Journal of Physical Anthropology* 101 (1996): 569–70, www.physanth.org/association/position-statements/biological-aspects-of-race.

29. American Anthropological Association, "Statement on 'Race,'" May 17, 1998, www.aaanet.org/stmts/racepp.htm.

CHAPTER 6: SOCIETIES AND INSTITUTIONS

1. Norbert Elias, *The Germans: Power Struggles and the Development of Habitus in the Nineteenth and Twentieth Centuries* (New York: Columbia University Press, 1996), 18–19.

2. Douglass C. North, *Understanding the Process of Economic Change* (Princeton, NJ: Princeton University Press, 2005), 99.

3. Nicholas Wade, *The Faith Instinct: How Religion Evolved and Why It Endures* (New York: Penguin Press, 2010), 124–43.

4. Napoleon A. Chagnon, "Life Histories, Blood Revenge, and Warfare in a Tribal Population," *Science* 239, no. 4843 (Feb. 28, 1988): 985–92.

5. Robert L. Carneiro, "A Theory of the Origin of the State," *Science* 169, no. 3947 (Aug. 21, 1970): 733–38.

6. Francis Fukuyama, *The Origins of Political Order: From Prehuman Times to the French Revolution* (New York: Farrar, Straus & Giroux, 2011), vol. 1, p. 48.

7. Ibid., 99.

8. "The Book of Lord Shang," Wikipedia, http://en.wikipedia.org/wiki/The_Book_of_Lord_Shang.

9. Fukuyama, *Origins of Political Order,* 421.

10. Ibid., 14.

11. Daron Acemoğlu and James A. Robinson, *Why Nations Fail: The Origins of Power, Prosperity, and Poverty* (New York: Crown, 2012), 398.
12. Ibid., 364.


CHAPTER 7: THE RECASTING OF HUMAN NATURE

1. Thomas Sowell, *Conquests and Cultures: An International History* (New York: Basic Books, 1999), 329.
2. Kenneth Pomeranz, *The Great Divergence: China, Europe, and the Making of the Modern World Economy* (Princeton, NJ: Princeton University Press, 2000), 3.
3. Gregory Clark, *A Farewell to Alms: A Brief Economic History of the World* (Princeton, NJ: Princeton University Press 2007), 127.
4. Ibid., 179.
5. Ibid., 234.
6. Nicholas Wade, *Before the Dawn: Recovering the Lost History of Our Ancestors* (New York: Penguin Press, 2007), 112.
7. Clark, *Farewell to Alms*, 259.
8. Ibid., 245.
9. Gregory Clark, "The Indicted and the Wealthy: Surnames, Reproductive Success, Genetic Selection and Social Class in Pre-Industrial England," Jan. 19, 2009, www.econ.ucdavis.edu/faculty/gclark/Farewell%20to%20Alms/Clark%20-Surnames.pdf.
10. Ron Unz, "How Social Darwinism Made Modern China: A Thousand Years of Meritocracy Shaped the Middle Kingdom," *The American Conservative*, Mar. 11, 2013, www.theamericanconservative.com/articles/how-social-darwinism-made-modern-china-248.
11. Toby E. Huff, *The Rise of Early Modern Science: Islam, China, and the West*, 2d ed. (New York: Cambridge University Press, 2003), 282.
12. Marta Mirazón Lahr, *The Evolution of Modern Human Diversity: A Study of Cranial Variation* (Cambridge, UK: Cambridge University Press, 1996), 263.
13. Marta Mirazón Lahr and Richard V. S. Wright, "The Question of Robusticity and the Relationship Between Cranial Size and Shape in *Homo sapiens*," *Journal of Human Evolution* 31, no. 2 (Aug. 1996): 157–91.
14. Richard Wrangham, interview, Edge.org, Feb. 2, 2002.
15. Norbert Elias, *The Civilizing Process: Sociogenetic and Psychogenetic Investigations* (Oxford, UK: Blackwell, 1994), 167.
16. Steven Pinker, *The Better Angels of Our Nature: Why Violence Has Declined* (New York: Viking, 2011), 48–50.
17. Ibid., 60–63.
18. Ibid., 149.
19. Ibid., 613.

20. Ibid., 614.
21. Jonathan Gibbons, ed., *2011 Global Study on Homicide: Trends, Context, Data* (Vienna: United Nations Office on Drugs and Crime, 2010).
22. Philip Carl Salzman, *Culture and Conflict in the Middle East* (Amherst, NY: Humanity Books, 2008), 184.
23. *Arab Human Development Report 2009: Challenges to Human Security in the Arab Countries* (New York: United Nations Development Programme, Regional Bureau for Arab States, 2009), 9.
24. Ibid., 193.
25. Martin Meredith, *The Fate of Africa: A History of Fifty Years of Independence* (New York: PublicAffairs, 2005), 682.
26. Richard Dowden, *Africa: Altered States, Ordinary Miracles* (New York: PublicAffairs, 2009), 535.
27. Shantayanan Devarajan and Wolfgang Fengler, "Africa's Economic Boom: Why the Pessimists and the Optimists Are Both Right," *Foreign Affairs*, May–June 2013, 68–81.
28. Clark, *Farewell to Alms*, 259–71.
29. Pomeranz, *Great Divergence*, 297.
30. Daron Acemoğlu and James A. Robinson, *Why Nations Fail: The Origins of Power, Prosperity, and Poverty* (New York: Crown, 2012), 73.
31. Lawrence E. Harrison and Samuel P. Huntington, eds., *Culture Matters: How Values Shape Human Progress* (New York: Basic Books, 2000), xiii.
32. Jeffrey Sachs, "Notes on a New Sociology of Economic Development," in Harrison and Huntington, *Culture Matters,* 29–43 (41–42 cited).
33. Nathan Glazer, "Disaggregating Culture," in Harrison and Huntington, *Culture Matters,* 219–31 (220–21 cited).
34. Daniel Etounga-Manguelle, "Does Africa Need a Cultural Adjustment Program?" in Harrison and Huntington, *Culture Matters,* 65–77.
35. Lawrence E. Harrison, *The Central Liberal Truth: How Politics Can Change a Culture and Save It from Itself* (Oxford, UK: Oxford University Press, 2006), 1.
36. Thomas Sowell, *Migrations and Cultures: A World View* (New York: Basic Books, 1996), 118.
37. Ibid., 192.
38. Ibid., 219.
39. Sowell, *Conquests and Cultures,* 330.
40. Sowell, *Migrations and Cultures,* 226.
41. Ibid., 57.
42. Christopher F. Chabris et al., "Most Reported Genetic Associations with General Intelligence Are Probably False Positives," *Psychological Science* 20, no. 10 (Sept. 24, 2012): 1–10.
43. Richard Lynn and Tatu Vanhanen, *IQ and Global Inequality* (Augusta, GA: Washington Summit, 2006), 238–39.

44. Ibid., 2.
45. Ibid., 277.
46. Ibid., 281.
47. Acemoğlu and Robinson, *Why Nations Fail,* 48.
48. Ibid., 238.
49. Ibid., 454.
50. Ibid., 211.
51. Ibid., 427.

CHAPTER 8: JEWISH ADAPTATIONS

1. Gertrude Himmelfarb, *The People of the Book: Philosemitism in England, from Cromwell to Churchill* (New York: Encounter Books, 2011), 3.
2. Charles Murray, "Jewish Genius," *Commentary,* Apr. 2007, 29–35.
3. Melvin Konner, *Unsettled: An Anthropology of the Jews* (New York: Viking Compass, 2003), 199.
4. Harry Ostrer, *Legacy: A Genetic History of the Jewish People* (New York: Oxford University Press, 2012), 92–93.
5. Anna C. Need, Dalia Kasparavičiūtè, Elizabeth T. Cirulli and David B. Goldstein, "A Genome-Wide Genetic Signature of Jewish Ancestry Perfectly Separates Individuals with and without Full Jewish Ancestry in a Large Random Sample of European Americans," *Genome Biology* 10, Issue 1, Article R7, 2009.
6. Gregory Cochran, Jason Hardy, and Henry Harpending, "Natural History of Ashkenazi Intelligence," *Journal of Biosocial Science* 38, no. 5 (Sept. 2006): 659–93.
7. Maristella Botticini and Zvi Eckstein, *The Chosen Few: How Education Shaped Jewish History, 70–1492* (Princeton, NJ: Princeton University Press, 2012), 109.
8. Ibid., 193.
9. Ibid., 267.
10. Konner, *Unsettled,* 189.
11. Neil Risch et al., "Geographic Distribution of Disease Mutations in the Ashkenazi Jewish Population Supports Genetic Drift over Selection," *American Journal of Human Genetics* 72, no. 4 (Apr. 2003): 812–22.
12. See, for instance, Nicholas Wade, *The Faith Instinct: How Religion Evolved and Why It Endures* (New York: Penguin Press, 2010), 157–72.
13. Botticini and Eckstein, *Chosen Few,* 150.
14. Jerry Z. Muller, *Capitalism and the Jews* (Princeton, NJ: Princeton University Press, 2010), 88.

CHAPTER 9: THE RISE OF THE WEST

1. William H. McNeill, *A World History* (New York: Oxford University Press, 1967), 295.
2. Victor Davis Hanson, *Carnage and Culture: Landmark Battles in the Rise to Western Power* (New York: Random House, 2001), 5.
3. Niall Ferguson, *Civilization: The West and the Rest* (London: Allen Lane, 2011), 18.
4. Toby E. Huff, *Intellectual Curiosity and the Scientific Revolution: A Global Perspective* (Cambridge, UK: Cambridge University Press, 2011), 126.
5. Ibid., 133.
6. Quoted in ibid., 110.
7. Jared Diamond, *Guns, Germs, and Steel: The Fates of Human Societies* (New York: Norton, 1997), 25.
8. Ibid., 21
9. IQ for Papua New Guinea is 83, compared with the European normalized score of 100. Richard Lynn and Tatu Vanhanen, *IQ and Global Inequality* (Augusta, GA: Washington Summit, 2006), 146. If Diamond has in mind some more appropriate measure of intelligence, he does not cite it.
10. Mark Elvin, *The Pattern of the Chinese Past* (Palo Alto, CA: Stanford University Press, 1973), 297–98, quoted in David S. Landes, *The Wealth and Poverty of Nations: Why Some Are So Rich and Some So Poor* (New York: Norton, 1998), 55.
11. Ferguson, *Civilization,* 13.
12. Ibid., 256–57.
13. Eric Jones, *The European Miracle: Environments, Economies, and Geopolitics in the History of Europe and Asia* (Cambridge, UK: Cambridge University Press, 2003), 61.
14. Ibid., 106.
15. Quoted in ibid., 153.
16. Jones, *European Miracle,* 61.
17. Huff, *Intellectual Curiosity,* 128.
18. Timur Kuran, *The Long Divergence: How Islamic Law Held Back the Middle East* (Princeton, NJ: Princeton University Press, 2011), 281.
19. Toby E. Huff, *The Rise of Early Modern Science: Islam, China, and the West* (Cambridge, UK: Cambridge University Press, 2003), 47.
20. Ibid., 321.
21. Ibid., 10.
22. David S. Landes, *The Wealth and Poverty of Nations: Why Some Are So Rich and Some So Poor* (New York: Norton, 1998), 56.
23. Lecture in 1755, quoted in Dugald Stewart, "Account of the Life and Writings of Adam Smith LL.D.," *Transactions of the Royal Society of*

*Edinburgh,* Jan. 21 and Mar. 18, 1793, section 4, repr. in *Collected Works of Dugald Stewart,* ed. William Hamilton (Edinburgh: Thomas Constable, 1854), vol. 10, 1–98.

24. Landes, *Wealth and Poverty,* 516.

CHAPTER 10: EVOLUTIONARY PERSPECTIVES ON RACE

1. This image is derived from an idea of Richard Dawkins that relates human and chimp ancestry.
2. Niall Ferguson, *Civilization: The West and the Rest* (London: Allen Lane, 2011), 322.

# INDEX